CHEMISTRY
and the
TECHNOLOGICAL
BACKLASH

CHEMISTRY
and the
TECHNOLOGICAL
BACKLASH

JAMES L. PYLE

Department of Chemistry
Miami University
Oxford, Ohio

Prentice-Hall, Inc., Englewood Cliffs, New Jersey

Library of Congress Cataloging in Publication Data

PYLE, JAMES L
   Chemistry and the technological backlash.

   Includes bibliographical references.
   1. Chemistry.   2. Technology—Social aspects.
I. Title.
QD33.P99   540   73-16073
ISBN 0-13-129528-4
ISBN 0-13-129510-1 (pbk.)

Permission for the publication
herein of Sadtler Standard Spectra ®
has been granted,
and all rights are reserved,
by Sadtler Research Laboratories, Inc.

10 9 8 7 6 5 4 3 2 1

Printed in the United States of America

PRENTICE-HALL INTERNATIONAL, INC., *London*
PRENTICE-HALL OF AUSTRALIA, PTY. LTD., *Sydney*
PRENTICE-HALL OF CANADA, LTD., *Toronto*
PRENTICE-HALL OF INDIA PRIVATE LIMITED, *New Delhi*
PRENTICE-HALL OF JAPAN, INC., *Tokyo*

To my wife, Betsy

Contents

Preface   xi

PART I

THE TECHNOLOGICAL BACKLASH   1

Introduction   1

The Technological Backlash, 3   Comments on Writing Organic Structures, 5
Chemical Orientations and "Some Chemical Comments", 8
Suggested Reading Section, 9

PART II
THE ENVIRONMENT WITHOUT   11

1 | The Energy Problem   13

*Chemical Orientation, 13   Perspective on Energy, 13   Energy from
the Sun, 15   Plants and Photosynthesis, 18   Carbohydrate Metabolism in
Animals, 20   Heat, Work, and Energy in the Biosphere, 23   The Fate of
the Plant, 25   Fossil Fuels and Man, 30   Alternatives to Fossil Fuels, 31
A Side Issue in the Energy Problem—Strip Mining, 34   New Directions in
Fossil Fuel Technology, 35   The Effect of Increasing Atmospheric CO$_2$, 36
Nuclear Energy, 39   Is There an "Energy Crunch" in Our Future?,   46
A Special Energy Problem—The Automobile, 47   Photochemical Smog, 53
Power Plants and Air Pollution, 60   Suggested Reading, 61*

2 | Pollution of Our Natural Waters by Nutrients   63

*Chemical Orientation, 63   Introduction, 63   The Role of Nitrogen, 64
Phosphorus Compounds as Nutrients, 66   Eutrophication: Ecological
Obesity, 68   The Solutions to the Eutrophication Problem: Getting at the
Sources, 73   The Chemistry and Ecology of Inorganic Fertilizers, 75
The Strange Story of Methemoglobinemia, 78   Soap and Detergents: The
Schizophrenic Molecules, 80   Synthetic Detergents, 86   Phosphates in
Detergents, 89   Sewage Treatment—The "Flush-It-and-Forget-It"
Syndrome, 94   Suggested Reading, 95*

3 | The Organic Chemicals Industries   97

*Chemical Orientation, 97   Petroleum—The Foundation of the Organic
Chemicals Industries, 99   Chemical Middlemen—The Chemical Process
Industry, 106   The Pulp and Paper Industry, 115   The Food Process
Industries, 122   Summary, 124   Suggested Reading, 126*

4 | Competition to Man from Other Species —
The Pesticide Problem   127

*Chemical Orientation, 127   Perspective on Pesticides, 127   Other
Insecticides, 133   Difficulties in the Biosphere with Chlorinated
Hydrocarbons, 136   Alternatives to Chlorinated Hydrocarbons, 142
Alternatives to Chemical Insecticides, 144   Chemical Communication
in Insects, 146   Juvenile Hormones and Insect Control, 150   Summary, 151
Fungicides, Rodenticides, and Herbicides, 151   Suggested Reading, 154*

5 | The Metallic Elements and the Environment   **157**

Chemical Orientation, 157   Introduction, 157   Iron and Steel, 158 Nonferrous Metals, 162   The Depletion of Metal Resources, 167   Toxicity of Heavy Metals, 169   The Role of Metal Ions in Metabolism, 170 Lead—An Old Killer in New Guise, 173   Mercury—The Perfect Pesticide for Man?, 176   Beryllium—A Toxicological Mystery—and Other Metallic Culprits, 183   Suggested Reading, 185

6 | An Untapped Resource—Solid Waste   **186**

The Old Problem, 186   Some New Solutions, 188   Packaging—A Separate Problem in Solid Waste Disposal, 192   Suggested Reading, 193

7 | The Chemical Detectives:
Pollution and Analytical Chemistry   **194**

Chemical Orientation, 194   Analysis of Pesticides, 195   Analysis of Trace Metals by Spectroscopic Analysis, 201   Air Pollution Analysis and Monitoring, 205   Suggested Reading, 208

8 | Pollution and Politics   **209**

PART III
THE ENVIRONMENT WITHIN   **215**

9 | The Population Problem   **217**

Chemical Orientation, 217   "Multiply and Subdue", 217   The Chemistry of Contraception, 222   Contraception in the Future, 232   Summary, 235 Suggested Reading, 236

10 | Extra Chemicals in Our Food   **237**

A Survey of the Food Additives, 237   The Artificial Sweeteners, 241 MSG—The All-Purpose Flavor Additive, 243   Diethylstilbestrol, 244 Nitrites, 246   Suggested Reading, 247

11 | The Chemical Crutches—The Drug Problem   248

Chemical Orientation, 248   The Nervous System, 249   The Rollercoaster: Amphetamines and Barbiturates, 258   The World of Unreality, 267 America's Most Serious Drug Problem, 273   Marijuana, 280   White Death: The Opiates, 290   Epilogue, 297

PART IV
SCIENCE—THE ETHICAL
AND POLITICAL DILEMMA   299

12 | The Scientist —
The Ethical and Political Dilemma   301

Chemical Orientation, 301   Science and Social Impact, 301   Case Study 1: "Manhattan District, U.S. Corps of Engineers", 302   Case Study 2: CBW—Public Health in Reverse, 306   Case Study 3: The Molecular Basis of Genetics, 309   Case Study 4: The Chemistry of Consciousness and the Intellect, 322   The Science Establishment, 323

Glossary   335

Index   345

Preface

Thermodynamics did not develop
the steam engine,
it was the other way around.

Bryce Crawford
Professor of Chemistry
University of Minnesota

We have seen great changes in the last decade, both in chemistry and in the needs of students of chemistry. It follows then that the teaching and learning of chemistry must also change. I hope that *Chemistry and the Technological Backlash* will serve as one such new avenue for the benefit of both teacher and student.

There are two general premises for the book. The first is that a new relationship is developing in man's attitude toward the management of technological development. In the past century, growth of the applied sciences has been staggering. They have changed the life and experience of ordinary men all over the world to a degree that the most farsighted prophet of Lincoln's time could not have dreamed of. The benefits of this technology have always been clear—the prolongation of life, the relief of physical labor, rapid transportation and communication, the development of better materials, an increase in man's ease and comfort. The accompanying deficiences of modern life have been far from obvious until recently to most of us. Yet there are important drawbacks. Prolonging life contributes to the population explosion. New energy forms are inefficient. Our modern system of transportation not only produces pollution, but consumes resources in greater measure. New synthetic materials developed for a worthy purpose may present ecological or toxicological difficulties, or may be difficult to dispose of. A drug may produce an unanticipated side-effect. Thus, we find ourselves at an impasse. While we must deal on the one hand with an economy and tradition geared for technological innovation, there now exists also a new philosophy of anti-technology, rejecting out-of-hand not only innovation but many aspects of what Americans have come to regard as the good life.

And what of chemistry in this situation? There is no question that, in its pure and applied forms, chemistry has played a major role in the development of this technology. Indeed "better things for better living" remains a source of professional pride. But if we are proud of the accomplishments, we need also to accept some blame for the shortcomings. For pollution problems and ecological imbalances have resulted in large measure from misapplication of the molecular science, usually with the approval of chemists. But in this respect we must make the very firm distinction between the basic knowledge-seeking function of the pure science of chemistry and its applied aspect. Every chemist is committed to the maxim that man will benefit from more knowledge of nature at the level of atoms and molecules. Indeed it is surely true that our problems with technology stem from ignorance, not malevolence. These are no industrial barons or crafty scientists willing to inflict some new product or invention upon us. Technological advance has always arisen from what its developers perceived to be a public need. It has usually been greeted with general enthusiasm, and its developers hailed as humanitarians and heroes.

The problem has not been that we have known too much; it has been that we have known too little. The backlash of technology has resulted from our failure to understand the complexities of biological and other systems on both the large scale and at the molecular level. By learning more about the chemistry of the system, it is usually possible to follow through with the proper adjustments in the technological application. Moreover, this knowledge is necessary to enact changes needed to correct the situation. Federal auto emission legislation is a good example of this.

It is clear that sensitivity to these problems is necessary in the training of new chemists. It is equally certain that an appreciation of the impact of chemistry upon problems of our time is a necessity in the intellectual portfolio of any educated person. Thus many chemistry teachers have begun to sense that they must relate the central place of chemistry in understanding and dealing with many great issues of our time along with the excitement of basic research and chemical theory. There is no doubt that students realize this need as well. These realizations are the reason for this book.

It is the author's hope that teachers and students may use *Chemistry and the Technological Backlash* as a resource and supplementary book to any of the several basic texts. The book is appropriate for the nonscience major, and we hope he is not unduly challenged at times by what may seem to be chemical sophistication. It is also written for the chemistry major, though we hope he is not unduly bored by what he may see as chemical simplicity on occasion. We have designed the book to be flexible. A teacher may assign chapters for study in any order which might be convenient to his particular course. In developing the text in a course for nonscientists at Miami University, we have found it convenient to introduce the materials in the second and third quarters as we teach organic, inorganic, and bio-chemistry after an initial quarter of traditional freshman chemistry. It is also possible for this book to be used alone for a quarter or semester of work, assuming that the student has had a good chemistry background, or that he has at least one quarter of college chemistry. The book will also be useful for advanced high school students or for elementary courses in organic chemistry or biochemistry.

Assistance with this work came from many sources, including my colleagues at Miami University in the Department of Chemistry and in the Institute of Environmental Sciences. Special thanks go to the unstinting typing efforts of the departmental secretaries, Janet Mercer, Randi Klees, and Jean Anthony. The assistance of the staffs of the Miami University Library and Audio-Visual Service was invaluable. Many students at Miami University contributed significantly to the work; in particular the contributions of Sara Goslee, John Rolfes, and James Profitt are worthy of mention.

JAMES L. PYLE

PART I

THE TECHNOLOGICAL BACKLASH

Introduction

THE TECHNOLOGICAL BACKLASH

It is in the nature of man to explore, to learn, and to apply what he has learned. Lacking other distinctions as a species, he has used this factor to rise in only a few hundred generations from primitive hunter to the complex interlocking system that comprises modern society. He has progressed from explorer of rocks for use as crude tools to explorer of rocks to tell him how the moon was born. He has spread throughout the world using his ability to adapt himself to whatever conditions of climate and geography he finds.

He has turned this knack of adapting new discovery to his every need. To supply his food he has learned to increase agricultural production by using new varieties of plants and animals, by developing inorganic fertilizers to stimulate plant growth and pesticides to control competitors, and by irrigating arid farm regions. He has supplemented natural fibers and animal skins with synthetic fabrics that he has learned to make from petroleum. He has supplemented the energy of his food supply to his muscles with that derived from coal and petroleum and the nucleus of the atom, to supply heat and cooling of his house and to generate electricity to do a thousand things to ease his work.

By learning about his body and about his biological enemies, he has learned to control or prevent diseases and to use surgical or chemical means to cure himself when he is sick or injured. He has gained great skill in moving from place to place. In the time from Magellan to Gagarin, 1522 to 1961, he has increased the speed of circumnavigation of the earth by a factor of 20,000. He can make himself heard or seen around the world in an instant. He has learned to store information and to process it at great speed. He has refined metals and other substances from the earth's resources into materials that can be used to construct whatever he may wish.

Until recently, the road to the future seemed clear and relatively unhindered. We would continue with more and more development, thereby easing poverty and disease, relieving discomfort, controlling the climate, traveling faster, and generally making life easier and better. Except for a few, this line of thought was dispelled only recently. If an event was needed to lead us in new directions, it may well have been, ironically, one of the greatest technological advances in the history of man; the flight of Apollo 8 in December, 1968. It was a momentous event because it marked the first occasion when man had departed entirely from earth's gravitational field. It marked a new consciousness upon the whole earth by way of remarkable photographs of our planet. We began to see that the earth is finite in its resources and that we had developed careless habits of use of these resources. Our energy consumption had increased markedly, for example, and this had begun to place strains upon our supplies of coal, and especially petroleum and natural gas. Of equal moment was the fact that many activities had produced wastes, and these wastes were at best disrupting the beauty of nature and at worst were so seriously disrupting the natural ecological equilibrium that it appeared that man himself might be endangered. Indeed, such pollution was proved to be a danger to man in some instances. His success in trimming down his own death rate in turn was causing a sharp upsurge in his own number. This was further straining his capacity to deal with the resources problem. The problem turned within himself as pesticides, food additives, drugs, and other chemicals taken into his own body proved to be more complex in their action than has been assumed, causing side effects or potential problems.

In the view of some, technology* has created a monster possessing the capacity to destroy man, and the future seems to be more of the same. That the present has brought the destructive capability of thermonuclear devices of power beyond man's grasp is at best a mixed blessing and at worst the seed of our destruction. That future technology may enable man to inject his thinking and will into the processes of his own conception, gestation, birth, and education in ways far beyond his present capacity is equally a frightening

---

* This word, and every other word in the text that has been marked with an asterisk, is defined in a Glossary appearing at the back of the book.

thought to many. Some individuals have chosen to reject technology and have attempted to return to older ways. This attitude has reached even to the Senate of the United States as reflected in their rejection of funding of the supersonic transport aircraft, clearly a technological advance, on environmental grounds.

It is not the thesis of this book that technology is bad. It is the thesis that unrestrained technology is bad and that the total impact of a technological advance must be weighed carefully before it is implemented. If the environmental degradation or social disruption is too serious, then the advance cannot be permitted. It is an irony that as we rewrite ground rules for change, we must continue to rely upon the technological specialists who have been blamed for some of the trouble. A principal figure in this area is the chemist.

It is true that chemists have been key participants in the development of industrial technology and its resulting pollution, for the development of persistent toxic chemicals, and for the development of harmful food additives and drugs. Yet while this admission must be made, it must be followed in the same breath by the knowledge that the problem lies not in chemistry itself but in its application. While it is true that most chemists entered the field because they saw in it a way in which they could use their skills to better man's lot, there was beyond this a curiosity about nature. Chemistry is a scientific, not a technological, discipline. It is a way of learning about nature, not a way of using her. That way is at the level of atoms and molecules. By understanding nature at this level, the chemist believes that our knowledge of the world can materially increase, and the fruits of success can be realized. That the advances of man's recent past have occurred is testimony in large measure to the truth of this philosophy. The problem for the chemist, as well as other scientists and technologists, is to work in a new context that demands a thorough examination of all ramifications of an advance before we set off on a new course. This means that chemists must work with professionals in other fields such as economics, health, politics, education, and the general public to work out solutions to a far greater degree than in the past. By the same token it is important that those who are not chemists understand the nature of technological and environmental problems at the molecular level so that they can deal with them more effectively.

It is the hope of the author that both the chemist and the nonchemist will gain understanding of great problems of our time by considering them from the chemical point of view. Before we take up the issues, there are two brief comments about features of this book that should be made.

COMMENTS ON WRITING ORGANIC STRUCTURES

Many of the chemical substances of concern in the book are organic compounds derived from the element carbon. Many of these compounds are quite complex, and to save time various abbreviated forms of structure writing are used. Take, for example, the compound known as *isopentane*, molecular formula C_5H_{12}. The Lewis structure* showing all valence electrons for the compound is

$$
\begin{array}{cccc}
\text{H} & \text{H} & \text{H} & \text{H} \\
\ddot{\text{H}}:\ddot{\text{C}} & : & \ddot{\text{C}} & : & \ddot{\text{C}}:\ddot{\text{C}}:\text{H} \\
\text{H} & \text{H}:\ddot{\text{C}}:\text{H} & \text{H} & \text{H} \\
& \text{H}
\end{array}
$$

in which the dots represent bonding electrons. A simplifying approach is to substitute lines for bonding electron pairs.

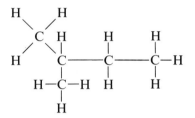

A further measure of simplification would be to write the structure without sharing the carbon–hydrogen bonds distinctly. Thus:

$$
\begin{array}{c}
\text{CH}_3 \\
\diagdown \\
\text{CH}-\text{CH}_2-\text{CH}_3 \\
| \\
\text{CH}_3
\end{array}
$$

or

$$(\text{CH}_3)_2-\text{CH}-\text{CH}_2-\text{CH}_3$$

An even simpler notation is a line structure in which even the individual carbon atoms are not shown.

This representation asks the reader to place a carbon at each line terminus and intersection and to fill out the hydrogens to the requirement that each carbon be tetravalent.*

In the nature of things these various structures will be used interchange-ably. When carbon–carbon double bonds or rings are formed, similar representations are used. Thus, cyclohexene, C_6H_{10},

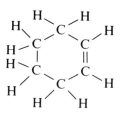

becomes

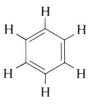

A special problem exists with benzene, C_6H_6, which is sometimes written

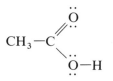

The problem is that all chemical evidence indicates that the double bonds are not present but have been incorporated into a new structural feature. For this reason we shall usually represent benzene by the notation

When atoms are present that contain nonbonding valence electrons,* our practice will be to show them as pairs of electrons. Thus, acetic acid may be represented as

$$CH_3-C \underset{\underset{\cdot\cdot}{O}-H}{\overset{\overset{\cdot\cdot}{O}}{\diagdown}}$$

although in some situations the carboxyl group* on the right side of the structure may be abbreviated

$$CH_3-CO_2H$$

In dealing with complex chemical structures it is important first to analyze the individual parts of the molecules. Consider estradiol, a female sex hormone, which will later be described in terms of its physiology.

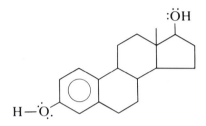

While the molecule is complex, we can sort out the complexity with ease. We note the presence of a benzene ring and three other rings all fused together. Three rings contain six carbons; the other contains five. We note also the presence of one methyl (CH_3) group designated by the solid line projecting upward from one of the ring junctions. We note also that two hydroxyl (OH) groups are present. The one on the five-membered ring is an alcohol;* the other located on the benzene ring makes it a phenol.* Thus, by inspection we gain a great deal of information about the compound.

CHEMICAL ORIENTATIONS AND "SOME CHEMICAL COMMENTS"

Before most chapters a brief paragraph describes what is called a *chemical orientation*. While the chemical essence of problems and issues are described, we do not usually take the time to describe the chemistry itself in detail. You may wish to do this by referring to a standard textbook or as directed by your instructor. The chemical orientation section may serve to indicate where you might usefully consult that textbook.

There also appear within the body of the text a few brief sections entitled "Some Chemical Comments." These comments contain a brief synopsis of a chemical principle or idea which is important to the discussions at that point. They are intended to be succinct and if you are using another text in conjunction with *Chemistry and the Technological Backlash*, you are likely to find more extended discussions there.

SUGGESTED READING SECTION

At the conclusion of each section a list entitled "Suggested Reading" is included. The idea here is to provide readers with additional sources for further exploration of a given topic. Some of the works are textbooks in chemistry or other fields. Some are references in periodicals that you may consult in your college library. The list of references is not intended to be inclusive, and you will usually find extensive bibliographies listed in most of the references.

Among these references several periodicals recur and a word about them is in order. *Science* is a periodical published weekly by the American Association for the Advancement of Science (AAAS).* It publishes original research papers in all sciences, review articles, and science news and comments. *Chemical and Engineering News* is also a weekly periodical published by the American Chemical Society (ACS).* While principally a news-magazine, it occasionally publishes work of a review nature. *Environmental Science and Technology* is also published by the ACS. *Scientific American* is a popular science periodical available to the general public. Its articles are written by scientists who are leaders in their fields. On the other hand, *American Scientist* is published by Sigma Xi, an honorary fraternity related to scientific research. *Environment* is published by the Scientists' Institute for Public Information, a group concerned with providing information to the American public on scientific and environmental concerns. A frequently cited reference is *Cleaning our Environment: The Chemical Basis for Action.* This book arose from a committee study completed on behalf of the American Chemical Society in 1969.

PART II

THE ENVIRONMENT WITHOUT

I

The Energy Problem

Energy is the ultimate currency of civilization.

Athelstan Spilhaus

Chemical Orientation   In the first part of this chapter we shall deal with the importance of energy in our society, with the nature of our basic dependence on energy from the sun and from the plants. We are concerned here also with the great quantities of energy that a modern society consumes. In this connection we shall consider the potential of nuclear energy.

For chemical orientation it is most important that you review sections of your chemistry text dealing with heat and energy in chemical reactions. You should also look over chapters concerning the structure, energy, and reactions of the atomic nucleus. There may be chapters or sections on carbohydrates, such as glucose, dealing with their chemical structure or metabolism or with the interactions of matter with light that may be useful to your study of this chapter. Finally, you should review the subject of oxidation and reduction reactions, in particular as they relate to organic compounds.

PERSPECTIVE ON ENERGY

Let us place ourselves in the role of the historian-philosopher for a moment and ask an interesting question. What kind of a yardstick could we use for measuring man's progression from his evolutionary beginning to the present? A case can be made that the best such measure would be his increasing ability to generate and utilize energy from sources beyond that of his own muscles. As he developed civilizations, man has come to use more and more energy, and the purposes he finds for it are also increasing. More sophisticated usage of energy is central to supplying food and shelter, to transportation and communication, to protection of man from his natural enemies, and to wage war against his own kind. It serves increasingly to enhance the productivity of each individual man and it makes possible new forms of pleasure and entertainment. In succession man has learned to use fire, to domesticate other animals, and then to develop simple machines for his agriculture and warfare. He has learned to use the wind and tides. The industrial revolution is said to begin with the steam engine and the modern era with the internal combustion engine. With the discovery of nuclear fission and fusion, the availability of energy is now so vast that many doubt whether man and his institutions can control its use properly.

In large measure, one can see the energy factor at work in our times by comparing energy usage and economic levels in different nations. Under-developed nations use only about 5% as much energy per capita as the United States. With 6% of the world's population, the United States uses 35% of the energy generated.

The problem facing an underdeveloped nation is to generate energy for industrial and agricultural growth, for transportation, and for consumer goods. This places additional burdens on the world's energy resources. One can also look at the question from the other direction: Can the United States justify such a disproportionate consumption of energy compared with the rest of mankind?

TABLE 1.1

Nation	Relative Per Capita Commercial Energy Consumption (1969)[1]
India	1
Brazil	2.5
Japan	14.7
Italy	12.6
Australia	26
USSR	22.7
USA	55.8

[1] Taken from Earl Cook, "The Flow of Energy in an Industrial Society," *Scientific American*, **224**(3), 142 (1971).

In Table 1.1 we compare commercial power consumption in selected countries. Each figure is on a *per capita* basis and all countries are compared to India. Thus Australia uses 26 times as much energy *per capita* as India.

What are the consequences in the generation of this energy, and what are its chemical aspects? How does the use of so much energy affect the environment and the ecological balance of living things? In this section we strive to answer these questions and to suggest some of the problems arising now and in the future from the utilization of vast amounts of energy by man.

The logical place to begin is to consider how nature provided for the energy of man and other animals. In the natural design, a pattern emerges that is ingenious in its simplicity on the one hand and yet so complex in its specific chemistry that it is still not understood in all its detail. In the simplicity there was a trick; two complementary systems were designed, each operating on a chemical reaction that was the reverse of that used by the other. Thus the products of each chemical reaction become the reactants for the other reaction. Most important is that the reaction cycle provides for entrance of energy from outside, from the sun. The chemical reactions are the means for energy transfer. The two complementary reactions are centered in the two great divisions of living things, the plant and animal kingdoms.

In order for plants and animals to exist and flourish, they must have access to the fundamental components of the universe—matter and energy. These components must be in the appropriate form to be utilized. From their environments they must extract substances which can be incorporated into their physical structure and which can be used as energy sources. These also must be available at the appropriate time and in proper concentration.

Our first concern is with this natural order, in particular as it relates to the *Energy Problem.* When that has been developed, we need then to consider the impact of modern man and how his energy needs in the postindustrial revolution era complicate and disturb the biological equilibrium.

ENERGY FROM THE SUN

Nearly all the earth's energy originates or originated from the sun. It arises from a nuclear reaction taking place in the mass of the sun whereby the element hydrogen is transformed to helium. In equation form this reaction is

$$4_1^1H \longrightarrow {}_2^4He + 2_{+1}^0\beta + energy$$

The immense heat of the sun, of the order of 1 million degrees centigrade, is required to cause the four hydrogen nuclei to coalesce into a single helium nuclei. Two positrons are emitted from the nucleus. The energy released from this process is enormously larger than in the conventional energy-releasing processes such as the burning of coal. One pound of hydrogen releases energy comparable to 10,000 tons of coal, about 10^{11} (100 billion) kcal (kilocalories),* upon fusion.

This large amount of energy is emitted from the sun into space in all directions in the form of radiant energy or waves traveling at the speed of light, 3.0×10^{10} centimeters per second. The waves given off have different *frequencies** or *wavelengths.** Some have relatively high frequencies, which means correspondingly short wavelengths. Others have lower frequencies or longer wavelengths. They extend through the entire electromagnetic* spectrum from short ultraviolet waves through the visible spectrum into the lower-frequency infrared and microwave region.

Some Chemical Comments

Three waves of differing wavelengths are shown in Fig. 1.1. Suppose arbitrarily that the center wave has a wavelength of 5,000 Å (angstroms*) (5.000×10^{-5} cm). The wavelength has the symbol λ and is so indicated on the wave (Fig. 1.1). The wavelength of the upper wave is half that of the center wave, or 2.500×10^{-5} cm; the lower wave is twice as great, or 10.000×10^{-5} cm. Since all the waves travel through space at the same speed, more of the shorter wavelengths will pass a given point per second. Thus the shorter wavelength has a higher *frequency* given in cycles per second or *hertz.** This can be calculated easily.

$$\text{Frequency (hertz)} = \frac{\text{speed of light}}{\text{wavelength}}$$

or

$$v = \frac{c}{\lambda}$$

where c is 3.00×10^{10} cm/sec. The frequencies of our three waves is given to the right in Fig. 1.1. Each wave is in a different region of the spectrum. The longest is in the infrared, the middle wave is in the visible region, and the shortest wave is in the ultraviolet region of the spectrum.

A small portion of the energy coming from the sun encounters the planet earth, $8\frac{1}{2}$ minutes later. This incoming energy is approximately 20 kilocalories per minute per square meter of the earth's cross section or a total of 2.56×10^{15} kcal reaching the outer reaches of the earth's atmosphere every

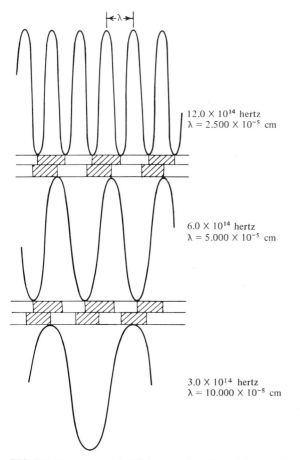

12.0×10^{14} hertz
$\lambda = 2.500 \times 10^{-5}$ cm

6.0×10^{14} hertz
$\lambda = 5.000 \times 10^{-5}$ cm

3.0×10^{14} hertz
$\lambda = 10.000 \times 10^{-5}$ cm

FIG. 1.1 Three waves of differing wavelengths and frequencies.

minute. Of this quantity about 35% is directly reflected back into outer space from the surface of the earth or from particles in the atmosphere. All the rest is absorbed by molecules in the atmosphere or on the surface. In the atmosphere much of the absorption of the energy waves is by water and carbon dioxide molecules. This energy is converted to heat and is radiated back into space by the earth. About 43% of the sun's energy reaching the earth is thus reflected after atmospheric absorption. This heat radiation is in the longer wavelength infrared region.

Of the total incoming radiation, then, about 78% is lost into space. The rest is stored in one form or another on the earth. Most enters the hydrologic cycle, in which the radiation energy is stored in the form of heat in water in the seas. This energy melts ice, evaporates, or simply warms water. Energy thus stored may be released as the processes reverse. These changes control climatic and weather conditions leading to precipitation, currents, storms, and other important factors that in effect move energy from one point to another on the earth.

Of the total energy that originally reached the earth, a tiny fraction, 600 billion kcal/min or 0.002 % comes in contact with green plants and is stored by them by a process called *photosynthesis.** It is this process that is central to nearly all aspects of the energy problem, and we shall now consider it in detail.

PLANTS AND PHOTOSYNTHESIS

From a chemical point of view, photosynthesis effects the conversion of CO_2 to more complex carbon compounds. The process is referred to by biologists as CO_2-fixing, which means that CO_2 is taken from the atmosphere and "fixed" in the plant, forming part of its cellular structure.

The process takes place in the cells of green leaves, in a nodular section called the *chloroplast.* Here the green pigment, chlorophyll, acts as a catalyst to initiate the photosynthetic process. Chlorophyll's ability to function in this way is dependent on molecular properties. The ability of chlorophyll to absorb the energy of the incoming light is what initiates photosynthesis.

This energy-absorbing ability is something that all molecules have. Energy absorption is accompanied by a change of residence of an electron. Electrons lie in *orbitals** or *energy levels** in both atoms and molecules. Of those energy levels available, the electrons will prefer to lie in those of lowest energy, and the atom or molecule is then said to be in its *ground state.* Some of the lowest unoccupied orbitals will be close in energy to some of the highest occupied orbitals. That means that not much energy is needed to promote

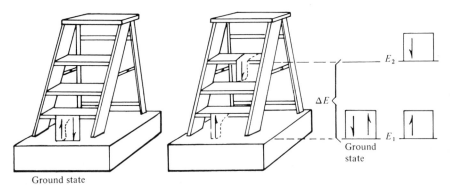

FIG. 1.2 Electronic energy levels may be viewed as analogous to steps on a ladder in that there are only certain levels permitted for the electron to occupy and specific energy difference between these levels.

or excite an electron into the higher energy level. The resultant molecule after energy absorption is in an *excited state*.* Many such transitions occur in the visible or in the ultraviolet regions of the spectrum, which means that sunlight radiating through the atmosphere can cause this excitation process to occur.

Energy levels are not unlike a stepladder, in that only certain levels are "allowed" and the energy difference between them is definite or quantized.

While the sequence of steps is not known in detail, especially the role played by the excited state molecule of chlorophyll, the result of several steps finds chlorophyll returned to its ground state and carbon dioxide reduced to stable natural carbohydrates, the class of organic compounds that contains sugars, starches, and cellulose. O_2 is a second product. The reaction sequence can be pictured as

Ground state chlorophyll + light $\longrightarrow$ excited state chlorophyll

CO_2 + excited state chlorophyll

$\qquad \longrightarrow$ carbohydrates + O_2 + ground state chlorophyll

A balanced summary of the synthesis of glucose, the most common sugar, in the form of a chemical equation is

$$6CO_2 + 6H_2O \xrightarrow[\text{light}]{\text{chlorophyll}} C_6H_{12}O_6 + 6O_2$$

The reaction is a heat-absorbing or endothermic process. The absorption of 673 kcal of energy accompanies the formation of each mole of glucose.

The reaction may be generalized for the production of any other carbohydrates or for cellulose, the principal component of plant fibrous material. Cellulose is a polymer consisting of many repeating glucose structural units linked together. The combination of two glucose molecules to form part of the cellulose is accompanied by loss of one water molecule, as a glycosidic linkage* forms between them.

Glucose occurs mostly in a cyclic structure with five carbons and one oxygen forming the ring. Four hydroxyl groups and a hydroxymethylene (CH_2OH) group are linked to the ring as shown (Fig. 1.3), and hydrogen atoms complete the structure. In forming the polymeric structure of cellulose, two glucose molecules come together, by splitting out a molecule of water, and forming an oxygen bridge between two rings.

Various sugars similar to glucose but differing in subtle structural ways can be synthesized in the plant by slightly different routes. By proper combinations to form larger structures, they constitute the structural material of plants and fill other roles. Principally in the form of starches, they also serve

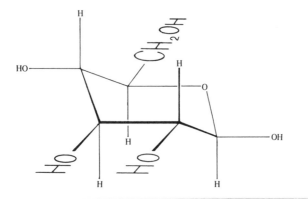

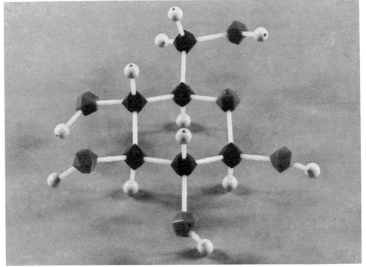

FIG. 1.3 Structure and photograph of one form of glucose known as β-D-(+)-glucose. The black and gray atoms are carbon and oxygen respectively; the white balls are hydrogen.

as food, and thereby as energy sources, for animals. Two such substances are cellulose and starch, both glucose polymers differing from one another in how the glucose units are linked together. Their structures are shown in Figs. 1.4 and 1.5.

CARBOHYDRATE METABOLISM IN ANIMALS

In the animal organism, be it protozoan, insect, or man, the process of photosynthesis is in effect reversed. Carbohydrates are oxidized to CO_2 and water in the cell, with the release of the energy from the chemical storehouse, which

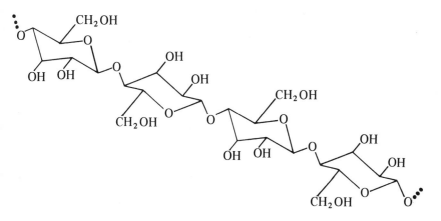

FIG. 1.4 Cellulose.

carbohydrates represent, to the surroundings—the cell. The polymeric glucoses such as starch are first broken down to glucose itself. This is accomplished in the higher animals by enzymes in the digestive tract. The smaller glucose molecules are readily absorbed from the digestive tract and transmitted to the cells where the oxidation takes place.

The oxidation* is the reverse reaction of the photosynthesis; that is,

$$C_6H_{12}O_6 + 6O_2 \longrightarrow 6H_2O + 6CO_2$$

Since the changes in bonding are the critical factors in determining heats of reaction, and since the processes exactly reverse the overall bonding changes in the photosynthesis reaction, the heat of reaction itself is also the reverse

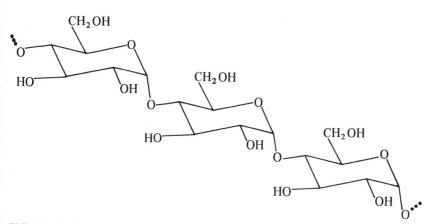

FIG. 1.5 Amylose, an important component of starch.

of photosynthesis. That is, $\Delta H = -673$ kcal for the oxidation of 1 mole of glucose. Thus the energy from the sun stored in the plant is now released. In the cells of animals the process is called *respiration*. This energy now available to animals is used for all the chemical reactions leading to physiological activity. Thus nerve stimulation, muscle contraction, glandular secretions, cell division, enzyme synthesis, and all other activities are fueled. Plants require respiration for their cellular activities in the same way that animals do.

What we have said so far is an oversimplification. Glucose is not oxidized to CO_2 and H_2O in one simple step. Many separate chemical reactions are required, and the breakdown occurs step by step with glucose being converted to CO_2 through a series of intermediates. The energy released is stored in another molecule, adenosine triphosphate, ATP.

The reaction of energy release specifically occurring in the cell is the loss of one phosphate group from ATP.

$$ATP + H_2O \longrightarrow \underset{\substack{\text{(adenosine} \\ \text{diphosphate)}}}{ADP} + H_3PO_4, \qquad \Delta H = -8.0 \text{ kcal}$$

This process, releasing energy, is how muscle contraction and other work is done in the cell. The reverse reaction, which is endothermic,

$$ADP + H_3PO_4 \longrightarrow ATP + H_2O, \qquad \Delta H = +8.0 \text{ kcal}$$

is how energy is stored. The energy released through the glucose oxidation sequences is taken up by ATP formation in several different steps. Each mole of glucose oxidized produces 38 moles of ATP on its way to becoming 6 moles of CO_2. The total storable energy then is

38 moles ATP/mole glucose × 8.0 kcal stored/mole ATP

$$= 304 \text{ kcal stored/mole glucose}$$

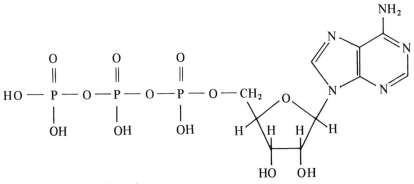

FIG. 1.6 Adenosine triphosphate.

The efficiency of the process is

$$\frac{304 \text{ kcal of energy stored in ATP}}{673 \text{ kcal released by glucose oxidation}} = 0.44 \text{ or } 44\%$$

Thus not all the energy is released in the form of useful work. Only 44% of the energy of the glucose molecule may be used in the respiration process. The rest is given off as heat in the course of the reaction sequences. This loss of heat always is observed in energy-releasing chemical processes.

HEAT, WORK, AND ENERGY IN THE BIOSPHERE

We have given thought now to the individual processes of photosynthesis and respiration. Since we are concerned about energy in this section, we need to focus on the energetics aspects of these reactions. The importance of the interrelationship of these processes is their role in energy transfer from the sun to plants and then to animals. We need to examine the interrelationship in the light of the science of *thermodynamics*, the science that studies how energy in the form of heat and work is involved in a chemical system.

What happens to the energy that the sun imparts to a plant in photosynthesis? We know that from the principles of conservation of mass and energy that the energy is not lost but has instead assumed some new form. This must mean that *something* has changed. When we look carefully at the plant, we find that we cannot account for a change on the basis of temperature change or other changes that have to do with the physical properties of the system. The only change is in the molecules. The energy change has happened because of a chemical reaction. The energy content of the products is higher than that of the reactants. The sunlight has been converted into *potential energy** of chemical bonds.

In other words the potential energy of the glucose and O_2 molecules formed in photosynthesis is greater than that of the CO_2 and water reactants. This can be so because the chemical bonding changes. The chemical bond is thus an energy storehouse. We can also understand from this how heat is given off or absorbed in a chemical reaction. It happens because bonds of different potential energies are broken and formed. If the potential energies of the products are lower than the reactants, some potential energy is converted to heat and given off to the surroundings. This is the exothermic reaction.* If the potential energy of the products is greater than the reactants, some of the energy must be taken up by the products from the surroundings. This is the endothermic process.*

There is another factor that is involved besides chemical changes and that is the capacity of a system to perform mechanical work. The most common case involves gases. A gas in expanding must push back the atmosphere. It does work and gives up potential energy. The reverse happens when a contraction of a gas occurs. This change in the energy of the system is usually small compared with potential energy changes involved in the chemical change. We shall neglect them here.

The First Law of Thermodynamics deals with the exchange of energy as we move from one situation to another. Its primary thrust is the idea of conservation of energy. As we move from reactant to product, energy changes its forms but it is not created or destroyed. In an endothermic reaction, heat energy from the surroundings is converted to potential energy in chemical bonds. This happens with photosynthesis.

A further ramification of the conservation of energy principle is that the energy of a certain system will be the same regardless of the origin of the system. One mole of CO_2 gas at 25°C and 1 atmosphere pressure represents such a system. Whether it formed from respiration, the sublimation of dry ice, the reaction of carbonate minerals in acid, or a physical separation from air, the energy situation is the same. From this standpoint two reverse processes such as photosynthesis and respiration form a cycle, the net result of which is zero energy change. As we go from CO_2/H_2O to glucose/O_2 and back to CO_2/H_2O, the energy absorbed by the one process is the same as that released by the other.

It further follows that what happens to a plant has no long-run impact on energy changes. There is only so much energy stored in the plant, and that energy is released slowly or quickly, now or later, but the end result is the same. In a forest fire the energy is released quickly and totally in the form of heat. If the plant becomes part of a fossil fuel deposit, the energy is stored for a very long period but its eventual return to the atmosphere in the form of CO_2/H_2O products with loss of energy is inevitable.

Besides the energetic changes, another important thermodynamic aspect is the concept of degrees of *order*, and *disorder* or *entropy* in chemical systems. It is possible to identify different degrees of order and disorder. A highly symmetric crystal in which each molecule occupies a specific lattice position with respect to its neighbors is a highly ordered state. A gas, where all the molecules move in a chaotic, random manner, is an example of a highly disordered state. Large molecules are more highly ordered than small ones, because more atoms are in fixed positions with respect to the other atoms.

We can see that as photosynthesis and respiration use reverse processes in the energetic sense, they are also reverse in terms of the change in entropy of the system. In photosynthesis where large (therefore ordered) molecules are generated from small gaseous (therefore disordered) molecules, the order increases in the reaction. In respiration the disorder increases. In terms of entropy, photosynthesis gives a negative change; respiration, a positive one.

THE FATE OF THE PLANT

In essence, we can see a complex equilibrium at work in nature involving carbon in its various oxidation states. This equilibrium, shown in Fig. 1.7, illustrates the components of this equilibrium. First we have the interactions between plants and animals and between atmospheric O_2 and CO_2 that we have discussed.

There are three common ways in which a plant meets its end and thereby releases its energy stored by photosynthesis. The most likely of these is that the plant dies and is slowly decomposed by the oxygen in the air. The cellulose comprising its structure and the other molecular constituents are oxidized, mainly to CO_2 and H_2O. Another possibility is that the same decomposition process takes place rapidly, as in a forest fire. Or the plant could be ingested by an animal and in the process either converted into the physical substance of the cells of the animal or metabolized to CO_2 and H_2O by respiration.

It is quite clear that all these routes lead sooner or later to the same molecular result: the carbohydrates present are broken down into CO_2 and H_2O, and the prototype equation may be thought to be the same for all of these. $(CH_2O)_n$ may be taken to represent any carbohydrate.

$$(CH_2O)_n + nO_2 \longrightarrow nCO_2 + nH_2O + \text{energy}$$

There is an interaction with the inorganic world, too. For instance, CO_2, in dissolving in water, finds itself in equilibrium with bicarbonate and carbonate ions and with calcium ions. This process is shown in these reactions.

$$CO_2 + 2H_2O \rightleftharpoons HCO_3^- + H_3O^+$$
$$HCO_3^- + H_2O \rightleftharpoons CO_3^{2-} + H_3O^+$$
$$Ca^{2+} + CO_3^{2-} \rightleftharpoons CaCO_3(s)$$

Thus CO_2 is lost from the biological world in the formation of limestone, $CaCO_3$. This process is also subject to reversal, however, by slow dissolving of limestone and other carbonate rocks that can revert back to CO_2, evaporate into the atmosphere, and reenter the biological chain. Carbonate ion is absorbed from the system by animals who utilize it in forming shells, coral, and the like.

Another possibility for the departure of matter from the equilibrium would happen if a plant were to be subject to decomposition in the absence of oxygen. This happens to a tree when it falls into a swamp and is covered over. It also occurs when algae and sea plankton settle at the bottom in the sea.

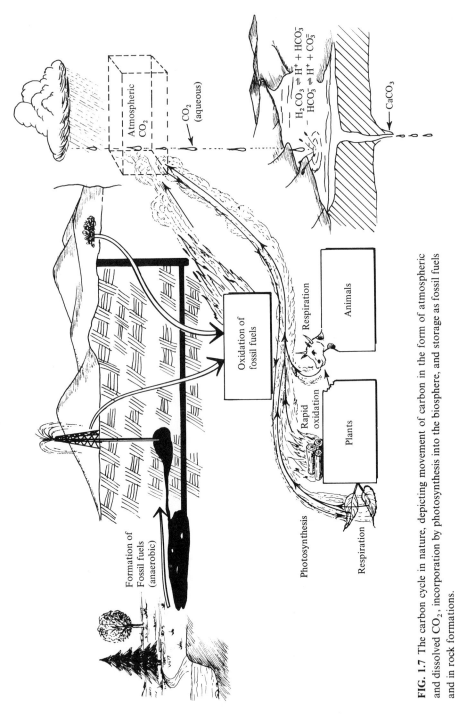

FIG. 1.7 The carbon cycle in nature, depicting movement of carbon in the form of atmospheric and dissolved CO_2, incorporation by photosynthesis into the biosphere, and storage as fossil fuels and in rock formations.

In such a situation the decomposition of the plant occurs by reduction. The products are hydrocarbons or elemental carbon, depending on conditions and length of time. We can write an equation to express this process, again in general terms:

$$\text{Carbohydrates } (CH_2O)_n \longrightarrow \text{hydrocarbons } (C_nH_{2n+2})$$

$$\text{Carbohydrates } (CH_2O)_n \longrightarrow \text{elemental carbon } (nC)$$

Some Chemical Comments

The formula C_nH_{2n+2} is a general formula representing a class of organic compounds called hydrocarbons or alkanes. The simplest member is methane, CH_4 ($n = 1$). The carbon atom is bonded to the 4 hydrogens by having two electrons in a covalent bond. For ethane, C_2H_6 ($n = 2$), the carbons form a covalent bond between them. Each carbon completes its four bonds by bonding to 3 hydrogens. As the number of carbon atoms increases, many possible isomeric arrangements become possible. With only 10 carbons, 75 structural arrangements are possible. The structures of some representative simple hydrocarbons are shown in Figs. 1.8 through 1.12.

The lower hydrocarbons with 1 to 3 carbons are gases at room temperature. Methane constitutes the principal component for natural gas. Petroleum is constituted by those hydrocarbons from the butanes to molecules

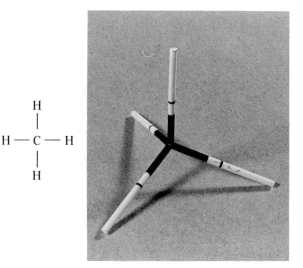

$$H - C - H \quad \overset{H}{\underset{H}{|}}$$

FIG. 1.8 Methane, CH_4.

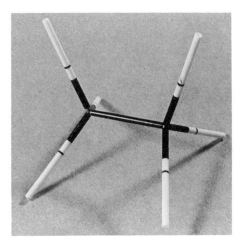

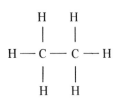

FIG. 1.9 Ethane, C_2H_6.

containing as many as 100 carbons. Coal is elemental carbon, although many hydrocarbons are present. Due to their common origin as the geological remains of plant decomposition, petroleum, natural gas, and coal are known as *fossil fuels.*

The net effect of the process of fossil fuel formation is to remove plants (and CO_2) from the dynamic biological system. The probability of the return of this matter to the active biological system over a long span of geological time is good. A swamp may dry up and the surface may erode, exposing coal deposits. Methane may form and build up sufficient pressure to force an opening through which escaping gas returns to atmosphere and is oxidized. An earthquake or drastic shift in sea level may expose a deposit to the atmosphere. These processes are very random and very slow, and one would indeed expect that the processes of formation and oxidation of fossil fuels,

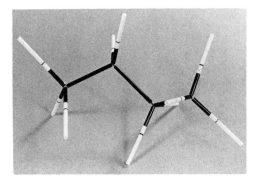

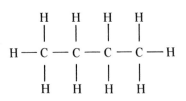

FIG. 1.10 *n*-Butane, C_4H_{10}.

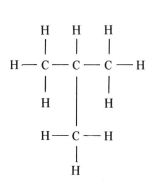

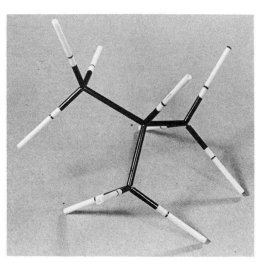

FIG. 1.11 Isobutane, C_4H_{10}.

both occurring slowly, would reach an equilibrium of their own so that the numbers of carbon atoms available to plants would change very little.

The equilibrium of the carbon cycle on earth is one to which life has adapted. Significant changes in conditions such as CO_2 level, the rate of fossil fuel formation, oxidation, or any other factor could be seriously disruptive to life on earth. Whether or not such a change is occurring is a topic to be considered later.

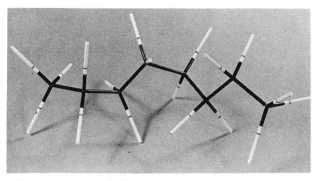

FIG. 1.12 n-Octane, C_8H_{18}.

FOSSIL FUELS AND MAN

Until the nineteenth century man had depended on the trees of the forests for those fuel needs that he required beyond those provided by his own metabolism and by that of his domestic animals. With depletion of wood supplies in populated regions and the discovery of extensive coal and petroleum deposits, man turned to these as fuels. They were cheap and abundant. The availability of this energy became a principal spur in accelerating industrial development. The trickle of the industrial revolution became a torrent.

Man's power to utilize the energy of the exothermic chemical reaction of fossil fuel combustion was thus magnified greatly. This became a curse as well as a blessing because of other products of the reaction, some of which adversely affect the environment when they are released in large quantity. Energy-producing reactions must be handled properly. They must be understood in terms of their chemistry, in terms of their ecological impact, and in terms of how we develop institutions to oversee their use.

Some salient facts illustrate what has occurred in the use of fossil fuels. In 1860, world production of coal and lignite was about 175 million tons; in 1910, 1,300 million tons; in 1970, 3,300 million tons, increasing by a factor of 18 in only 1 century. In 1860, world production of crude oil was nominal. By 1910 it reached 363 million barrels; by 1940, 2.2 billion barrels; by 1960,

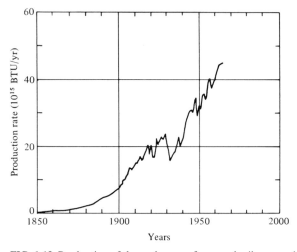

FIG. 1.13 Production of thermal energy from coal, oil, gas, and water power in the United States, exclusive of Alaska. Note the decrease in energy usage that accompanied the great depression of the 1930s. (From *Resources and Man*, M. King Hubbert, ed., W. H. Freeman and Co., San Francisco, Calif., 1969, p. 165. By permission of the National Academy of Sciences, Washington, D.C.).

7.6 billion barrels. The annual world consumption has now reached the 15-billion barrel level. The use of petroleum has increased by 7 times in the last 30 years. Natural gas usage has increased similarly in the same period.

If one looks at the overall picture of the United States by combining these sources of thermal energy, one sees the data in Fig. 1.13. Note that the energy derived from coal, oil, gas, and water power shows a steep curve that, except for the period of the great depression in the 1930's, is rather continuous and smooth. Overall we see that mankind uses five times as much energy per year as in 1900 from fossil fuels.

It is obvious that we cannot continue at such rates of increase along this curve into the future, using fossil fuels and other conventional sources of power, for a simple reason. Their supply is finite and therefore limited. Projections of exhaustion of the fossil fuel resources are rather complex. Estimates of how the rate of usage will increase from present levels must be made. The rate of discovery of new fields must also be considered, and sometimes the discovery of large new deposits such as the petroleum in Alaska's North Slope in the late 1960's can change the estimates. These problems notwithstanding, geologists and engineers have been able to predict the limits of these resources. For petroleum and natural gas, the prediction is about a century; for coal, 300 to 400 years. When these time estimates are thought of in the context of history, we see that our future reliance on fossil fuels is going to have to decrease. This is especially true of the hydrocarbon fuels. Alternatives must be found to avert an energy crisis.

What are the possible solutions? One is to find acceptable alternative energy sources that can supply energy equivalent to that now consumed by fossil fuel combustion and can also meet expanded future needs. Another is to limit per capita energy usage. A third would be to devise some way to prepare hydrocarbons rapidly from plant sources. Some microorganisms which will reduce plant cellulose and sugars to hydrocarbons are known, and some studies have been undertaken to check the feasibility of such processes which in effect produce petroleum artificially. All alternatives must be subjected to close examination.

ALTERNATIVES TO FOSSIL FUELS

The thought of alternatives requires that we first establish present levels of usage from all the energy sources. Electrical generation of power in the United States has consumed fuel in various categories in recent decades, as shown in Table 1.2. The data in summary shows a sharp increase in petroleum and gas consumptions, while coal has remained constant and the proportion of hydroelectric usage has dropped. The impact of nuclear power on the total market has not yet been of significance. It is ironic that we seem to be

TABLE 1.2
PERCENTAGE DISTRIBUTION OF SOURCES OF ELECTRICAL POWER

Year	1940[1]	1950[1]	1960[1]	1970[2]	(1980)[2]
Coal	54.6	47.1	53.6	55.0	45
Petroleum and gas	12.1	23.8	27.1	26.6	19
Hydroelectric	33.4	29.2	19.0	15.1	11
Nuclear	—	—	negligible	3.3	25

[1] U.S. Bureau of the Census, *Statistical Abstracts of the United States*; 1969, 40th ed., Washington, D.C., 1969, p. 512.
[2] *Chemical and Engineering News*, **48**(25), 16 (1970).

relying increasingly on those fossil fuels that are the least abundant. This dependence cannot continue into the future, as the projected data for 1980 show. The only alternative source that can pick up the load, and which is presently technologically feasible, is nuclear energy. That is the reason for extensive plans for nuclear power plant construction by power companies in the 1970's, which in turn has led in many instances to controversy with environmentalists.

The Federal Power Commission has predicted power usage will show a fourfold increase from 1,529 to 5,852 billion kw-hr (kilowatt hours) per year between 1970 and 1990. The ramification of this increase upon our fossil fuel and hydroelectric reserves will be clearly traumatic. As these resources decrease, the cost of power will increase, assuming that the distribution of energy sources remains about the same as the present for the various energy sources. This assumption is clearly not acceptable, and two developments could change it. One would be if the recovery efficiency of usable fuels from fossil fuels were to increase, and this is not likely to be significant because it is already fairly good. The other would be if hydroelectric, nuclear, and other sources were to replace fossil fuels.

In connection with the latter possibility, one could increase hydroelectric reserves by the development of more stream sites and the construction of more hydroelectric dams. All accessible sites are already in use and many of these are subject to the development of sediment over the years, which seriously limits their capacity for power generation. Also, the construction of a dam means the disruption of a natural stream site with scenic, recreational, and conservational values of its own. This loss may frequently be too great to justify the benefits in power generation. At most, assuming ideal conditions and the essentially complete development of all water power reserves, the total would constitute only a threefold increase in power from hydroelectric operations. A more reasonable estimate, taking economic limitations into account, might be a twofold increase in hydroelectric capacity. With the projected fivefold increase in power usage, the percentage coming from hydroelectric sources should thus decline further in the last decades of the twentieth century.

Other possibilities have been explored. The use of the energy of moving water from tides has been proposed, but the prospects for this are limited by the number of sites available. The heat generated by the interior of the earth has been proposed as a potential energy resource. This *geothermal* energy source suffers from the same limited potential and technical problems as the tidal sources. There simply is not enough energy easily available to sustain a level of power usage approximating that sustained by fossil fuels and water power.

The problems of direct use of solar energy do not deal with limited resources. As we have said, the energy reaching the earth from the sun is vast. Only a tiny fraction is trapped by plants in photosynthesis thereby being converted to energy sources available to man. Farrington Daniels, in his book *Direct Use of the Sun's Energy*, estimates the thermal power which reaches the earth from the sun to be 17.7×10^4 billion kilowatts* which represents 1.55×10^9 billion kilowatt hours per year. Given the world's total generating capacity to be 3.8×10^3 billion kilowatt hours, we see that the solar input is about 100,000 times that generated by all the electrical-generating plants on earth. The problem then becomes one of trapping the energy and converting it to other forms. Various means have been utilized to collect solar energy, but the essence of the problem is technological. For instance, sunlight falling on a square base 6.5 km would be required to match the 1,000-Mw (megawatt) power output of most modern power stations. There are few places with so much space not used for other purposes, although many desert areas would qualify. Further, the problem is compounded by cloud cover and seasonal changes and by the fact that the generators could not be used at night. Assembling enough solar batteries for this area is a complex task.

The problems to be solved are substantial, but solar energy could serve as a useful adjunct to classical methods. For instance, the use of solar energy could be used to charge batteries that then could be discharged to provide energy for heating and cooling a home. When the efficiency dropped, in cloudy weather, a conventional power source could be tapped. Such approaches, whether economically competitive or not, would lead to a reduction in fossil fuel consumption, though probably not a substantial one.

Scientists are sharply divided on the question of the future of solar energy. Some believe that it is the answer to man's energy need. Others believe that the technological problems are too difficult to enable solar energy ever to compete on economically even terms with conventional sources. There seems general agreement on two points: that we have not done enough to develop the use of solar energy in some applications and that further research will enable us to find more useful means of exploiting solar energy. The possibility that collecting stations for solar energy could be set up in outer space, with energy being transported to earth by laser beam, is an intriguing suggestion. As problems and controversy have arisen respecting nuclear energy, more and more attention is being directed toward exploitation of solar energy. It is thus possible that its development may increase rapidly, and it may indeed be a principal energy by the end of the century.

A SIDE ISSUE IN THE ENERGY PROBLEM —STRIP MINING

In taking the coal from the earth, man has long had to pay a high human price. The miner had a most dangerous calling as he ventured beneath the earth. Injury and loss of life to accident, cave-in, or explosion was a constant fear. Then economic and technological developments changed this. Automation displaced many miners, first beneath the ground, then above it. The strip mine, in which the soil cover is removed and the entire coal seam is dug out by machine (see Fig. 1.14) is an increasing practice. It remains one of the most controversial environmental issues of our time. A visit to a strip mine

FIG. 1.14 Strip mining. The large machine is used to remove the soil cover, thereby exposing the coal seam. (Photo by Bill Garlow, *Journal Herald*, Dayton, Ohio.)

location immediately reveals some aspects of the problem. By removing vegetation, the area loses its beauty and, worse still, it becomes subject to severe soil erosion. The problem is partially alleviated if mine operators replant the area, although years are required to restore a forest to something approaching its original condition. The trend in recent years has been for stricter state laws requiring restoration.

There is a more subtle problem connected with strip mining, and to some degree in deep mining as well. The villain is the sulfur present in coal, occurring principally as iron pyrite, FeS_2. When a coal seam is exposed to air, a reaction occurs oxidizing the sulfur to sulfuric acid. Microorganisms may play a role in the oxidation.

$$2FeS_2 + 2H_2O + 7O_2 \longrightarrow 2FeSO_4 + 2H_2SO_4$$

This acid drains into lakes and streams and causes severe ecological damage by killing wildlife. This in turn destroys recreational opportunities.

The reason for strip mining is that it is cheaper to operate large mechanical equipment than to hire many miners. The advantage that this brings means cheaper energy for the consumer but, at the same time, this factor must somehow be balanced against loss of employment in the minefields, and the disruption this has brought to areas such as Appalachia, against the safety hazards of deep mining and against the ecological, economic, and aesthetic damage to the land and streams.

NEW DIRECTIONS IN FOSSIL FUEL TECHNOLOGY

The need to clean our environment while supplying our energy needs principally through fossil fuels has left us in an ironic dilemma. The problem is that the "dirtiest" fuel, coal, is the most abundant; the "cleanest," natural gas, is the scarcest. One solution to the dilemma is to convert the abundant fuel into the scarce one, thus solving the pollution and supply problems with one stroke. The question is—how, especially when economic factors are taken into account.

At present, intensive research is being directed toward the development of coal gasification techniques. A typical experimental method is to grind the coal to a powder. Steam and oxygen are added to the coal at 1,000°C and reaction produces a mixture of carbon monoxide, hydrogen, and methane. They are formed from coal (elemental carbon) in these processes.

$$C + H_2O \longrightarrow CO + H_2$$
$$C + 2H_2 \longrightarrow CH_4$$

Tars and dusts, as well as sulfur oxides, are then removed from the gas. In the presence of a catalyst, some of the carbon monoxide is then converted to CO_2 by steam.

$$CO + H_2O \longrightarrow H_2 + CO_2$$

Hydrogen is now added in excess to convert CO to the desired product, methane. The methanation reaction is

$$CO + 3H_2 \longrightarrow H_2O + CH_4$$

Since the process occurs at high temperature and pressure, and this is expensive, the problem is to reduce the cost of the process. A strong point in favor of coal gasification is that the sulfur present in most coal in environmentally objectionable levels is removed. Thus both the particulate and sulfur oxide pollutants that cause problems in coal-burning operations are relieved in the gasification operation.

THE EFFECT OF INCREASING ATMOSPHERIC CO_2

The combustion of hydrocarbons in the automobile and in electric power generation produces CO_2 and water, in addition to energy. The production of water has no effect on the atmosphere because the amount produced is minimal compared to that already present. There is only a small amount of CO_2 under equilibrium conditions in the atmosphere. A balance has existed over time between the production of CO_2 by the natural oxidative processes of animal metabolism of sugars and oxidative plant decay on the one hand with photosynthesis and other processes that remove CO_2 from the atmosphere.

The combustion of fossil fuels in the last 100 years has changed the equilibrium condition. The production of CO_2 now exceeds the rate of removal. As a result, the CO_2 content of the atmosphere is increasing slowly. Since 1890 the concentration has increased from 290 to 320 ppm (parts per million). Figure 1.15 shows that a measureable increase has occurred since 1958. The increased CO_2 output has been compensated for in part by removal of some of the CO_2 into the oceans, either dissolved or in the form of carbonate, but the rate of this process has not increased. Can such a small change as 30 ppm exert any appreciable effect? Many scientists feel that it can.

The effect on plant life at first glance would seem to be favorable. The more CO_2 there is, the more photosynthesis should be facilitated, since CO_2 is a reactant in that process. Biologists have conducted experiments in which they have measured photosynthesis rates in atmospheres of higher CO_2

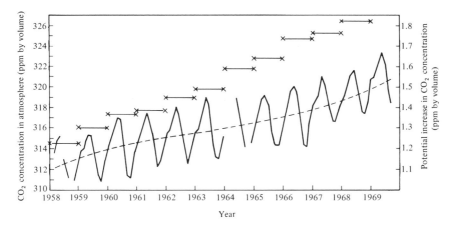

FIG. 1.15 This graph shows the measured increase in atmospheric CO_2 at Mauna Loa, Hawaii, between 1958 and 1969. Note that there is a seasonal variation, with a high point in late winter and a minimum in late fall. The dotted line depicts the yearly average and the straight lines indicate the yearly increase if all fossil fuels burned remained in the atmosphere. Similar increases have been recorded at several places in the world so the result is not a local one. (From *Man's Impact on the Global Environment*; *Report of the Study of Critical Environmental Problems*, MIT Press, Cambridge, Mass., 1970, p. 48.)

concentration. The results agree with our expectations. Photosynthesis and plant growth are enhanced; however, CO_2 also has a subtle and very adverse effect that relates to its ability to absorb energy in the infrared* region of the spectrum. Like almost all organic compounds, CO_2 absorbs energy in this region of the electromagnetic spectrum, the part of the spectrum that lies in a broad region centered at a wavelength of 15μ [a frequency of 2×10^7 hertz]. Much of the radiant energy from the sun when it strikes the earth is reflected immediately back into space. It so happens that the frequency of the CO_2 absorption is in the same region as the frequency of some of this reflecting radiation.

This means that some of this energy is absorbed by CO_2 molecules instead of continuing out of the earth's atmosphere and, when emitted by the molecule, the energy stays in the atmosphere. The consequence of this on a larger scale would be to increase the temperature of the atmosphere. A fourfold increase in CO_2 concentration is possible in the next 500 years at the present rate of fossil fuel consumption. If that increase takes place, we would expect an increase of 7°C in the average surface temperature of the earth. This seems small on first glance but the effects are far-reaching. At the higher temperature, the vapor pressure of water will be higher; hence, the humidity and extent of cloud cover would increase. This would lead in turn to dramatic changes in climate on the land. Further, this increase will have drastic effects on ecosystems. Some plants and animals, both terrestrial and marine, would not survive. Those species dependent on these for food would also be affected, creating a domino effect. Perhaps the most drastic change would come about because of increased melting of the polar ice caps. This would increase the level of the oceans, consequently inundating many coastal areas.

Is there any evidence that any of this is happening? There is certainly well-documented evidence of CO_2 increase in the atmosphere, as we have described it. There is also some evidence of a small increase in temperatures in the twentieth century both in the sea and in land areas. The problem is that any measurement of temperature changes must take into account that there are cyclic changes in temperature from other climatic factors that may be considered normal. Also, the amount of data on weather and temperatures before 1900 is sparse and thus it is difficult to know the normal temperatures over a longer period. The best conclusion at the moment is that there probably has not been enough increase in the concentration of CO_2 to effect significant differences, but scientists have enough confidence in the theory to feel that the process will be inevitable if the rate of increase of CO_2 in the atmosphere continues.

An ironic counter-theory to this temperature increase concept has been postulated. It involves the fact that the amount of particulate matter arising from many of man's activities is also increasing in the atmosphere. The expected effect of the particles is to reflect some of the sun's radiation out of the atmosphere, thereby acting to reduce the amount that reaches the earth's surface. This would decrease the temperature. Could these two effects cancel one another? No sane person would want to try that approach. It would be safer to attempt to correct the effects of both CO_2 and particulates in the atmosphere by reducing their levels.

One interesting aspect of the equilibrium may be seen by considering the relative amounts of carbon involved at various stages of the carbon cycle. While it is relatively easy to determine the amount of CO_2 in the atmosphere,

TABLE 1.3[1]

Atmospheric carbon	683×10^9 metric tons
Organic carbon—terrestrial	$1,580 \times 10^9$ metric tons
Organic carbon—oceanic	703×10^9 metric tons
Estimated dissolved carbon (as CO_2)	41.0×10^{12} metric tons
Annual fossil fuel combustion	15.4×10^9 metric tons

[1] Taken from *Man's Impact on the Global Environment*, report of the Study of Critical Environmental Problems (SCEP), The MIT Press, Cambridge, Mass. and London, England (1970), pp. 160–163.

that contained in the sea and in living plants and animals—the so-called *biomass*—is more difficult to measure accurately. Table 1.3 shows the best estimates for these values that research has produced.

We see that the quantity of fossil fuels presently burned each year is sufficient to equal present atmospheric CO_2 levels in about 44 years. Since fossil fuel combustion rates may increase threefold in the next 30 years, this increase will accelerate. Even though the sea presents a large potential reservoir for CO_2, increased rates of absorption of CO_2 into water are small because the turnover of deep seawater is slow. It is not thought that the removal of CO_2 in this manner is significant. Another factor to be considered is that the clearing of forests reduces the biomass, and thus photosynthesis, and also contributes to increases in atmospheric CO_2 through decomposition.

NUCLEAR ENERGY

Nuclear energy is the most reasonable alternative to fossil fuel power generation in the opinion of most experts. An assessment of its problems and potential, its hopes, and its hazards requires that we first examine what is known about atomic structure and the energy sources involved in the atom. We begin with some historical developments that led to our present concepts of atomic and nuclear structure.

Some Chemical Comments

Experiments by Michael Faraday and others in the nineteenth century demonstrated clearly that atoms contain electrical charges within their structure, and the neutral atom was conceived to consist of particles oppositely charged. The electrons* were negative; the protons* were positive. This model of the atom, accepted by most chemists and physicists at the turn of the century, was one that involved placing the particles into the atom in a diffuse way. Both the protons and electrons were thought to be distributed through the whole volume of the atom. Such a distribution of particles was thought most reasonable because it meant that there were no centers of high positive or negative charge. This model was known as the Thomson atom, named for the English scientist J. J. Thomson.

A crucial experiment was performed in 1910 by Sir Ernest Rutherford. The experiment was simple. A very thin gold foil sheet was bombarded by α-particles* traveling at a high velocity. Using scintillation-detecting devices that gave off a small amount of light when struck, the scattering of the particles as they passed through the gold foil was observed. It is possible to predict how the scattering would be affected in the Thomson atom. Since the α-particles are positively

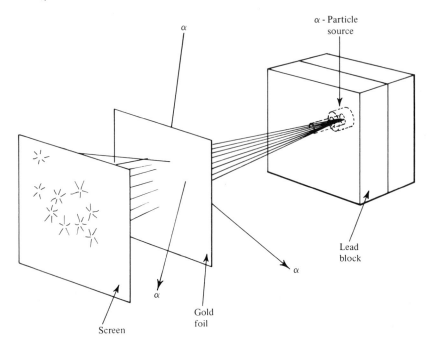

FIG. 1.16 Diagram of Rutherford's apparatus for studying the scattering of α-particles by gold foil.

charged, one would expect deflection because of repulsion when a particle happened to pass near a positive charge in an atom. Since the Thomson model required a diffuse distribution of the protons throughout the atomic volume, the expected result would be that many α-particles would be deflected, but that the degree or angle of deflection would be small because there was no center of high charge. Rutherford found that only a very small number of α-particles were deflected, which indicated that few particles were passing near a positive center; but the surprise was that some particles were deflected through large angles (see Fig. 1.16). This implied a large repulsion had taken place, which meant that a high positive charge center existed in the atom.

The divergence of these results from those expected led to the discarding of the Thomson model. The Rutherford model that took its place differed in two significant ways. The existence of a very small nucleus with a diameter of about 10^{-12} cm, compared with a diameter of 10^{-8} cm, for the whole atom was postulated. In this center was contained all the positive charges and nearly all the atomic mass. The existence of a very high charge density had accounted for the effect in the Rutherford experiment. Few particles were deflected because most did not pass close enough to the nucleus to be repelled at the high velocity at which the α-particles were traveling. Substantial deflections occurred when an occasional particle passed very near the nucleus. A direct hit on a nucleus would cause a deflection of the α-particle back toward the source, and indeed this was sometimes observed.

Thus our knowledge of a compact nucleus of high mass and high charge was born. The existence of isotopes was accounted for by postulation of the neutron as a third particle of the atom. An understanding of the diffuse negatively charged electrons interacting with the Rutherford nucleus but maintained in peripheral orbitals of definable energy then began to emerge. Upon development of quantum mechanics* in the 1920's this understanding was greatly advanced.

A serious question is still posed by the existence of the atomic nucleus. If the protons in the nucleus were as densely packed as Rutherford's experiment showed, what force existed that could overcome the enormous repulsion of their like positive charges? One part of the answer to this problem was found by close measurements of atomic masses and comparison of those masses with the masses of the constituent particles. In the smallest nucleus with more than one particle, a proton and a neutron are combined to form the nucleus of deuterium, an isotope of hydrogen. Since the respective masses of the two particles are independently known, a simple sum should give the mass of deuterium. Thus:

Mass of the proton	1.00729 atomic mass units* (amu)
Mass of the neutron	1.00866 atomic mass units
Expected mass of deuterium nucleus	2.01595 atomic mass units
Actual mass of deuterium nucleus	2.01355 atomic mass units
Mass loss	0.00240 atomic mass units

We are caught short by the fact that this sum is not the same as the true observed mass of the deuterium nucleus. Thus, a loss in mass has occurred by the formation of the nucleus from the particles.

When we express Δm, the weight loss, in grams, and c, the speed of light, in centimeters per second, the combined units from ΔE are gram-centimeters squared per second squared (g cm^2/sec^2), which is equivalent to the erg. To calculate ΔE in ergs, we convert the mass change to grams. To do this we divide by Avogadro's Number.

$$\frac{-0.00240 \text{ amu}}{\text{Deuterium nucleus}} \div 6.02 \times 10^{23} \frac{\text{amu}}{\text{g}}$$

$$= -3.95 \times 10^{-27} \frac{\text{g}}{\text{Deuterium nucleus}}$$

This value is Δm, taken as a negative number because of the loss of mass involved.

Now to calculate ΔE, using the famous equation, $\Delta E = \Delta mc^2$,

$$\Delta E = -3.95 \times 10^{-27} \text{ g} \times (2.9979 \times 10^{10} \text{ cm/sec})^2$$

$$= -35.5 \times 10^{-7} \text{ ergs or } -3.55 \times 10^{-6} \text{ ergs}$$

This is extremely small energy change, but remember that it is for a single deuterium nucleus. Some perspective can be gained by converting this number back to the molar level, using Avogadro's Number. In effect we are then calculating ΔE on the molar level for the nuclear reaction

$$_1^1\text{H} + _0^1 n \longrightarrow _1^2\text{H}$$

$$\Delta E = -3.549 \times 10^{-6} \text{ ergs/nucleus} \times 6.0225 \times 10^{23} \text{ D nuclei/mole}$$

$$= -2.14 \times 10^{18} \text{ ergs/mole}$$

By converting this value to kilocalories,

$$-2.14 \times 10^{18} \text{ ergs/mole} \times 2.390 \times 10^{-11} \text{ kcal/erg}$$

$$= -5.11 \times 10^7 \text{ kcal/mole of deuterium}$$

As a basic for comparison, recall that the value for the energy change upon combustion of a mole of glucose was 673 kcal. This is a rather large value as energy changes in chemical reactions go. Many values lie in the region of 0 to 200 kcal/mole. Yet the simple nuclear reaction,

$$_1^1\text{H} + _0^1 n \longrightarrow _1^2\text{H}$$

has a tremendously higher energy output. Thus the binding energy of the nucleus that the change in mass represents is many orders of magnitude stronger than analogous energies associated with chemical bonds in molecules. It is this huge energy potential that represents an attraction to man as he wrestles with the energy needs that sophisticated technology requires. By causing energy-releasing nuclear reactions to occur, this source can be tapped.

By the 1930's, understanding of atomic structure had begun to mature. An idea both exciting and ominous emerged based on two known facts: that nuclides could indeed change from one to another by natural radioactive decay processes and that these nuclear reactions could release vastly more energy than chemical changes. The natural consequence was the possibility that nuclear changes could be induced artificially and, if they could, that they could perhaps be controlled. The work of Otto Hahn succeeded in demonstrating that a very powerful nuclear reaction could be induced by bombardment of nuclei by the neutrons from the outside. Furthermore, in the case of the uranium isotope $^{235}_{92}U$, the decay itself led to emission of neutrons that could cause the process to continue in a chain reaction with the concomitant release of vast amounts of energy. This discovery led to the atomic bomb (better named the nuclear bomb).

There are numerous products of uranium fission, of which one is

$$^{235}_{92}U + ^{1}_{0}n \longrightarrow ^{139}_{56}Ba + ^{94}_{36}Kr + 3^{1}_{0}n$$

$$\Delta E = -4.7 \times 10^9 \text{ kcal/mole of } U^{235}$$

With the emission of three neutrons for each uranium atom bombarded, the process can continue as the product neutrons bombard other U^{235} nuclei until all the isotope has been consumed. This would be an uncontrolled process. If one could control this reaction, it should be possible to effect the conversion slowly enough that the energy involved can be absorbed, transmitted, and used. This can be done if only some of the product neutrons are allowed to react further, and others are prevented from doing so. The continuing chain reaction is attenuated, and the process is controlled.

The nuclear reactor is designed to accomplish this control. The nuclear reaction occurring in the reactor is the same spontaneous degradation of U^{235}, but the reaction is controlled by the fact that only a few of the neutrons produced strike new U^{235} nuclei. The others are absorbed by other nuclei present which do not then undergo further fission. The chain reaction is thus sustained but controlled. On the other hand there is the problem of too little U^{235} in nature. It is present only in small amounts, about 0.71% of naturally occurring uranium. By increasing the proportion of U^{235} in the sample, one can expect an increase in efficiency of the process, since the chances of the neutron hitting a U^{235} nucleus will be enhanced. If sufficient U^{235} nuclei are present, they will absorb the product neutrons and thus a chain is initiated. The chain will thus be sustained only if this *critical mass* of U^{235} is present. If too many neutrons are absorbed by other atoms, then the chain will not be extended. The first problem is to concentrate U^{235} from natural uranium to a high enough level to sustain the reaction.

Once the chain reaction starts, a balance must be struck between sustaining and controlling the reaction. This is done in atomic reactors by the use of control rods. These are sometimes constructed of cadmium, which absorbs neutrons without fission. As the number of neutrons increases, the control rod is inserted into the reaction core and its contact with the reacting fissile material is increased, until the system is brought to an equilibrium in which the production of power is both controlled and efficient.

The generation of power rises from the heat of the nuclear reaction. The core of the reactor is in contact with water or molten sodium metal that is heated by the exothermic nuclear reaction. This heat converts water to steam directly or indirectly, and this steam drives electrical generators. Then this water is cooled and recirculated to the reactor core in a closed system. Cooling is done by heat transfer to other water that is usually taken from a stream or lake and then released back to its origin. The increased temperature of the coolant water causes the problems of thermal pollution associated with nuclear power plants (see Fig. 1.17).

A modification in reactor technology is leading to the development of reactors known as *breeder reactors*. The modification has to do with the fact that one can convert the plentiful isotope U^{238} into a fissionable isotope U^{239} by neutron bombardment; i.e.,

Step. 1. Neutron bombardment:

$$^{238}_{92}U + ^{1}_{0}n \longrightarrow ^{239}_{92}U$$

Step 2. β-Emission:

$$^{239}_{92}U \longrightarrow ^{239}_{93}Np + ^{0}_{-1}\beta$$

Step 3. A second β-emission:

$$^{239}_{93}Np \longrightarrow ^{239}_{94}Pu + ^{0}_{-1}\beta$$

Plutonium^{239} undergoes fission under neutron bombardment similarly to U^{235}. The advantage of this system is that we are using an abundant and therefore less expensive uranium isotope as starting material. The problem of depletion of existing U^{235} resources is lessened. Since a gram of U^{238} would produce about 5×10^9 kcal by this route, about a million times more than that produced in the glucose combustion, the possibility of utilizing this energy in power generation was obvious from the discovery of the fission process.

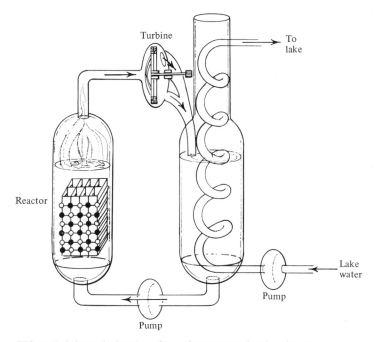

FIG. 1.17 Schematic drawing of a nuclear reactor showing the reactor core, power-generating turbines, and primary and secondary coolant.

As the potential in nuclear power generation is great, so are the problems, however. One of these is that control of the reaction requires vigilance, caution, and a high degree of instrumental reliability. It is imperative that instrumental and human monitoring and control systems work properly to keep the reactor under control and to prevent any large-scale radiation emission to the environment. Numerous safeguards are built into the reactor to prevent these eventualities. Nevertheless, it has not been the practice to construct reactors in heavily populated areas. To some degree this adds expense of transporting nuclear power from more remote areas to populated centers.

Then there is the problem of the other products of the nuclear reaction besides energy, namely the chemical products. These slowly decompose by natural radiation. Some of the isotopes have long enough half-lives that they will retain a high degree of radioactivity for many decades or even centuries. One must find a method and site of disposal that is expected to remain stable for hundreds of years. Disposal at sea is dangerous because leakage of the container could lead to a spread of the radioactive wastes over large areas of the sea, harming marine life. Disposal in a cave or remote desert area is another possibility, but dangers exist even there from earthquakes or leakage of wastes into underground water or to the atmosphere. Design of containers must be sufficiently sturdy and resistant to corrosion to last for a long period. The Atomic Energy Commission had considered disposing of radioactive wastes in an abandoned salt mine in central Kansas, but later abandoned the plan in the face of local objections.

Another major concern with reactors is thermal pollution. Not all the heat of the reactor can be converted to work in the turbine generators. Indeed this is the substance of the Second Law of Thermodynamics, as we have seen. Moreover, it is not as economical to recover all the heat generated in the reactor from the water that circulates through it. As a consequence the practice at many reactors is to release water used into streams and lakes at a temperature several degrees higher than the resident waters. This can lead to serious disruptions in the aquatic ecological situation, especially if there is little current flow to disperse the warm water. The use of cooling towers to release some of the energy into the atmosphere is one approach, although this form of heat transfer is not without problems, and it obviously adds to the expense. Another solution is to require that more of the energy be recovered in the power process. Presumably this would mean a somewhat higher cost per energy unit because it is now cheaper to disperse the heat. This price may be worth paying to assure a high energy conversion from the fuel and to avoid ecological side effects. The idea of using the warm water for use in agriculture where the heat could be used to enhance plant growth is another attractive approach.

IS THERE AN "ENERGY CRUNCH" IN OUR FUTURE?

There are several reasons to think so. In the long run we face fossil fuel shortages that force us to examine alternative energy sources. These alternatives, the breeder reactor, solar energy exploitation, and others, are not widely available or await further technical development. Meanwhile, our population increases and so does our per capita energy consumption. As our power consumption increases, the problem can only become worse.

In the short run two factors have combined to make the situation critical. One is the drastic decrease in the supply of natural gas. This has happened because of changes in law, reducing subsidies and depletion allowances to natural gas developers, which has decreased the growth of new gas wells. Another factor is the increasing criticism of new power plant construction, especially nuclear installations, on environmental grounds. This has meant changes in plans or plant sites and in delay. The situation presents itself as a classic in the environmental dilemma of our times between the need for more energy and the need to be conscious of safety, pollution, and conservation concerns in planning.

The energy crisis which became of concern in 1972 relates to the long-term questions of supply and demand which we have discussed, but there are other nonchemical factors of importance in the context of the immediate problem. Policies relating to the taxation of oil producers can stimulate or discourage domestic reserves exploitation and refinery construction. The use of the trans-Alaskan pipeline will increase supply but with environmental costs. Federal air quality standards affecting the automobile have led to some reduction in gas mileage in newer model autos because of pollution controls. Massive importation of foreign oil, which finds the United States, Japan, and Europe competing for the crude oil produced by the Arab states of the Middle East, raises new economic and political questions of the greatest importance for all these nations.

A SPECIAL ENERGY PROBLEM—THE AUTOMOBILE

Chemical Orientation   In the last section of "The Energy Problem" we are dealing with the automobile in connection with its status both as a consumer of energy resources and as an air pollution problem. Here, as in the first section, we shall be dealing with oxidation and reduction of both organic and inorganic compounds. You should also review the structure of molecular oxygen and read any sections of your text dealing with the subject of free radicals, molecules containing unpaired electrons.

The problem of the automobile and the utilization of energy is a special one. As a combuster of fossil fuels, it contributes in a major way to their depletion—about half the petroleum produced in the United States is processed into gasoline as the final product to be used by its 103 million motor vehicles. In addition, problems arise because the combustion is difficult both to sustain and to complete. Pollution of the air and other parts of the environment results as well.

The internal combustion engine derives its energy from the exothermic process of hydrocarbon oxidation or combustion. A prototype of the reaction would be the combustion of the hydrocarbon, *n*-octane, whose molecular formula is C_8H_{18}. Its structure consists of a chain of eight carbon atoms bonded together with the hydrogens bonded to the carbons in such a way that each carbon has four bonds. Its structure is shown in Fig. 1.12, p. 29. As in the previous oxidations, the products are CO_2 and H_2O. The balanced equation is

$$2C_8H_{18}(g) + 25O_2(g) \longrightarrow 16CO_2(g) + 18H_2O(g)$$

$$\Delta H = -2634 \text{ kcal (energy released by the reaction)}$$

FIG. 1.18 This is a typical situation in many American cities in that we see a heavy concentration of traffic on the inbound section of a freeway during the rush hour. Such traffic tie-ups, resulting from encouragement of automobile use and the decline of mass transportation systems, is a part of nearly every American city's experience. (Photo by Fred Straub, Staff Photographer, *Cincinnati Enquirer*.)

The combustion process occurs when the gasoline in a fine spray is mixed with air in the carburetor. This mixture is injected into the cylinder and there is ignited by a spark. The reaction, thus initiated, then takes place and in

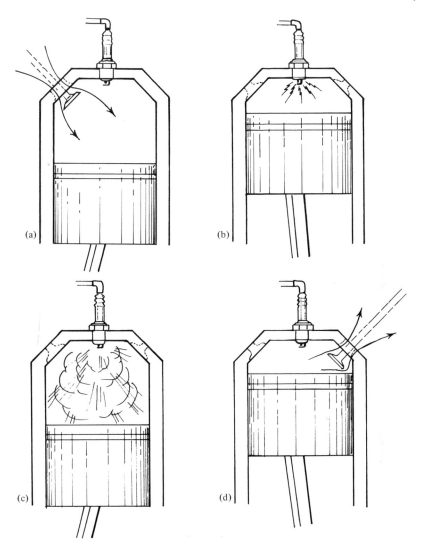

FIG. 1.19 Four steps in the operation of the internal combustion engine: (a) intake of fuel–air mixture, (b) compression of mixture, (c) ignition and combustion, and (d) exhaust of product gases.

the process pressure builds up in the cylinder. Note that 34 moles of gas are produced from 27, and this and the increased temperature are the reason for the pressure development. The expansion in the cylinder that the pressure produces provides the power for the automobile. The piston is pushed backward and this mechanical energy is transferred to the crankshaft of the automobile and thence into motion. The piston next recoils expelling the exhaust gases through an exhaust manifold. Then the cycle is repeated. Figure 1.19 shows the operation of the cylinder.

The properties of the fuel are naturally important in the efficient operation of the engine. Two problems, essentially opposite in character, arise. In high compression engines, there is a tendency for the gasoline to ignite before the piston reaches its maximum compression. This leads to knocking because the power stroke that follows is uneven and out of synchronization with the other cylinders in the engine. The opposite problem is incomplete combustion, which results in the presence of hydrocarbons and products of incomplete oxidation of hydrocarbons in the exhaust. Both problems have made great environmental impacts.

Several approaches have been used to combat the knocking problem. It was earlier observed that hydrocarbons having branched structures, like isobutane Fig. 1.11 (p. 29), rather than straight chains of carbon atoms tend to have better antiknocking qualities. Petroleum chemists thus set about devising methods to increase the amount of branched molecules in the gasoline. By treatment in the presence of solid catalysts at high temperatures some of the straight chain molecules can be converted to the branched analogs. Examples of the structure differences are the isomers n-octane and 2,2,3-trimethylpentane, both having the molecular formula C_8H_{18} (Fig. 1.20).

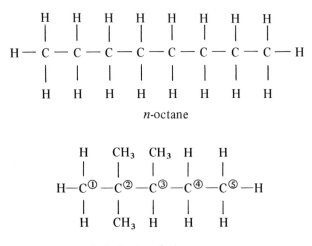

n-octane

2, 2, 3-trimethylpentane

FIG. 1.20 The structure of n-Octane and 2,2,3-Trimethylpentane (both C_8H_{18}). (Numbers indicate the carbon positions to which the methyl groups are bonded.)

The next development was the discovery that knocking properties of gasoline were improved by the addition of tetraethyl lead. The structure of this compound is

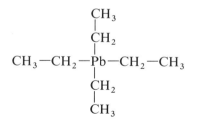

$$CH_3-CH_2-\underset{\underset{\underset{CH_3}{|}}{\underset{CH_2}{|}}}{\overset{\overset{\overset{CH_3}{|}}{\overset{CH_2}{|}}}{Pb}}-CH_2-CH_3$$

Like many other lead compounds, tetraethyl lead (TEL) is highly toxic, but since its concentration in the fuel is low, this has not been a serious concern. Most commercial leaded brands contain about 3 ml of the additive per gallon, a concentration that has not been found harmful under ordinary exposure conditions.

A more serious problem has to do with what happens to the lead as it passes through the cylinder and then through the exhaust system into the atmosphere. In the presence of oxygen under the conditions of high temperatures and pressures in the cylinder, the expected product should be the oxides of lead, particularly PbO. This oxide is a typical inorganic oxide with a high melting point (886°C). As a result of its low volatility, a good deal of the inorganic lead product deposited in the engine and proved troublesome. A solution that was developed has been to produce $PbBr_2$ instead of the oxide. This is accomplished by the presence of another additive, ethylene dibromide, which has the structure

$$\underset{\underset{Br}{|}}{CH_2}-\underset{\underset{Br}{|}}{CH_2}$$

On reaction with PbO, the lead bromide forms. It is more volatile, does not cause deposit problems, and is expelled in the exhaust. Lead bromide is extremely toxic. The effect of its elimination in the exhaust and its consequent presence in the environment is beginning to show some adverse effects in the view of many biologists. These are seen most strikingly in the plant life in some areas with heavy auto traffic where the concentrations of lead are high. It is clear that the concentration of lead compounds in the atmosphere is higher in populated areas, and no one questions that the principal source of this contamination is the tetraethyl lead in gasoline. The problem of lead toxicity will be discussed in a later section. The annual rate of consumption of lead in gasoline is now approaching half a billion pounds. As in many controversial environmental issues, there are several potential solutions. Each bears some cost to society. In the question of leaded gasoline the options are as follows:

1. We can continue to use tetraethyl lead in fuel. The question of harmful environmental effects has not been established beyond question, but if the fears prove to be true, we run the risk of harm upon the environment and upon ourselves that may persist for many years.

2. We can change the composition of fuel by improving the antiknock properties of fuels. One way to do this is to use other additives that may convey the same antiknock properties as TEL. Another is to improve the combustibility characteristics of the fuel by increasing its branched hydrocarbon characteristics. This is technically feasible, but it would add to the cost of fuel.

   New formulations of fuels appearing in recent years under "low lead" or "no lead" labels have contained increasing amounts of aromatic hydrocarbons. In particular, toluene is used. Aromatic hydrocarbons, compounds related to benzene, tend to burn a little less readily than other hydrocarbons. When the aromatic composition is increased, the fuel is less likely to fire too soon in the cylinder. On the other hand, there is some concern that if aromatics are present in exhaust, they may lead to new adverse health effects since some of them are carcinogenic. At present, there is little evidence to show auto exhaust as a significant source of these compounds compared with other sources.

3. We can change the fuel requirements of engines. Engines of lower compression capabilities could use fuels of lower knocking characteristics effectively, but at the cost of performance capabilities such as fast acceleration. The question is whether the public would accept such a change. In 1971 United States manufacturers decreased the compression ratio of the autos from 10 to 1 to 8.5 to 1. This decrease permits the use of fuel of somewhat lower antiknock characteristics, usually termed lower octane ratings; low-lead fuels perform adequately in this lower compression engine.

4. Another option is to turn to other types of engines. Steam engines have received attention. Here the combustion of fuel heats water or some other liquid, and this heat is converted to mechanical energy. The combustion is external to the piston and the mechanical action. The fuel requirements are far less rigorous. A major controversy is whether the steam engine is technically feasible and safe for mobile systems used in such great numbers.

   Among other engines being explored for use instead of the standard internal combustion engine is the Wankel engine, which uses a rotary compression system instead of the back-and-forth action of the standard piston. Its advantage is a lighter weight, and its use includes the possibility of including exhaust control equipment without increasing engine size. In fuel injection systems the fuel is burned twice, first as a rich fuel mixture and then as a lean mixture that enables the more complete combustion. This reduces fuel requirements and improves emission. Autos powered by electric batteries have also been brought forth as a solution, not only to the lead problem but also as a means of alleviating auto-caused air pollution. There are major problems with the electric automobile. Its range is limited because the batteries must be recharged. The lead necessary in the battery will increase the weight of the car drastically. Perhaps the most important disadvantage is that the stress on our already overloaded power-generating capacity would increase. Then the increase in air pollution from power plant operations would offset some of the decrease in auto emissions. We are still limited by the First Law of Thermodynamics. Energy cannot be generated from no energy. To transfer the energy from one form to another may only change the nature of the stress without improving the total situation.

5. There is a final option—to reduce our dependence on the car, especially in urban areas. This may mean legal or economic restrictions on cars, especially larger cars, together with encouragement of mass transit systems. This would require a major decision by many individuals to give up the convenience and efficiency of the car. The merits of bicycles and of the feet as transport media are being revived.

There are several choices to solve this problem, which is a nice situation in a way. The petroleum industry is choosing the option of elimination of TEL while retaining the antiknock properties with other additives. The auto industry seems fully committed to the internal combustion engine and is not seriously studying alternatives. The government itself has not moved against lead additives, but the fact that lead compounds clog pollution devices in exhausts and that the government is requiring more rigorous emission standards accomplishes this indirectly. The options of reducing automobile traffic in urban centers may become important later in the 1970s.

PHOTOCHEMICAL SMOG

The second problem of the internal combustion engine is that the process is not completed in the cylinder. Products other than CO_2 and H_2O form. They represent intermediate oxidation stages. One of these is carbon monoxide, CO. Other compounds representing still less complete oxidation are produced, and indeed some completely unreacted hydrocarbon molecules will be emitted. In addition to all of this, there is an undesirable side-reaction resulting from oxidation of nitrogen from the air.

All of these substances, CO, hydrocarbons, and nitrogen oxides, can play a deleterious role in the air. Together with the light they cause smog formation, called *photochemical smog* to denote the requirement that light be present to cause the chemical changes. In order for smog to develop, at least three conditions must be met. Two of the components that we previously mentioned, nitrogen oxides and hydrocarbons, must be present. Of the nitrogen oxides, nitrous oxide (NO) and nitrogen dioxide (NO_2) are present in the exhaust. Once emitted into the air, NO can be oxidized to NO_2 by atmospheric O_2.

The composition of hydrocarbons in the exhaust emission is very complex. Some of the hydrocarbon molecules may pass through the car cylinder unchanged. Some fragments, called *free radicals*,* form and these in turn will react quickly with O_2 or other substances in the air. A free radical is a substance that has an unpaired electron in its structure. Such a substance could form by the removal of a hydrogen atom from *n*-butane; for example,

$$CH_3-\underset{\underset{H}{|}}{CH}-CH_2-CH_3 \longrightarrow CH_3-\underset{\cdot}{CH}-CH_2-CH_3 + \cdot H$$

Since O_2 in its ground state is known to be a diradical (two unpaired electrons), one expects a reaction between the hydrocarbon radical and O_2 because the pairing of two radicals to form a bond is usually a facile process.

$$CH_3-CH-CH_2-CH_3 + :\overset{..}{\underset{\cdot}{O}}-\overset{..}{\underset{\cdot}{O}}: \longrightarrow CH_3-\underset{\underset{:\overset{..}{O}-\overset{..}{\underset{..}{O}}\cdot}{|}}{CH}-CH_2-CH_3$$

This new species is called a *peroxy radical.*

Other species derived from hydrocarbons may be thought of as intermediate oxidation steps between the hydrocarbons and CO_2. If we consider successive removal of hydrogen atoms or the addition of oxygen atoms to represent successive oxidations, we can devise the scale of changing oxidation as shown at the top of page 55.

The number appearing by each structure represents the oxidation state* of the indicated carbon atom.

One must not assume that this is the chemical route actually followed in the combustion, but all of these kinds of molecules are found in auto emissions. In particular the presence of aldehydes such as acetaldehyde, ketones (Fig. 1.21), and the unsaturated hydrocarbons such as ethylene is significant. The reason for their importance has to do with the ability of the unsaturated carbon–oxygen or carbon–carbon double bond, also known as the π-bond, to absorb light. This absorption of light promotes an electron from a ground state bonding orbital to a higher energy antibonding orbital.* Such excited states can undergo reactions that will not occur with the ground state molecule.

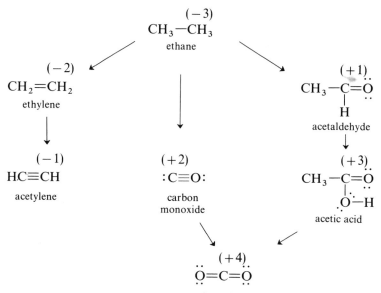

(-3)
$CH_3—CH_3$
ethane

(-2)
$CH_2=CH_2$
ethylene

$(+1)$
$CH_3—\overset{..}{\underset{H}{C}}=\overset{..}{\overset{..}{O}}$
acetaldehyde

(-1)
$HC\equiv CH$
acetylene

$(+2)$
$:C\equiv O:$
carbon
monoxide

$(+3)$
$CH_3—\overset{..}{\underset{\overset{..}{O}—H}{C}}=\overset{..}{O}$
acetic acid

$(+4)$
$\overset{..}{\underset{..}{O}}=C=\overset{..}{\underset{..}{O}}$
carbon dioxide

Understanding of the process by which hydrocarbons, their various oxidation products, nitrogen oxides, and atmospheric oxygen interact with light to form photochemical smog is incomplete, but chemists have succeeded in learning some of the details of the process. A scenario for the process reads as follows:

Under the influence of light, NO_2 is split into NO and atomic oxygen.

$$:\overset{..}{\underset{..}{O}}—\overset{..}{N}=\overset{..}{\underset{..}{O}}: \xrightarrow{\text{light}} \cdot N=\overset{..}{\underset{..}{O}}: + \cdot\overset{..}{\underset{..}{O}}\cdot$$

The NO shortly reoxidizes to NO_2 but the atomic oxygen is extremely reactive. It forms ozone, O_3, by reaction with normal molecular oxygen.

$$:\overset{..}{\underset{.}{O}}—\overset{..}{\underset{..}{O}}: + \cdot\overset{..}{\underset{..}{O}}\cdot \longrightarrow :O=\overset{..}{\underset{..}{O}}—\overset{..}{\underset{..}{O}}:$$

FIG. 1.21 A representative ketone, 2-butanone. A ketone shares with an aldehyde the presence of the carbonyl group $\overset{\diagdown}{\underset{\diagup}{C}}=\overset{..}{\underset{..}{O}}$. The difference is that the aldehyde has at least one hydrogen bonded to the carbonyl carbon. The ketone has two alkyl groups bonded to the carbonyl carbon.

$$CH_3—\overset{\underset{\parallel}{C}}{\underset{:O:}{}}—CH_2—CH_3$$

One of the interesting questions is the problem of ferreting out the role of an excited state of molecular oxygen, called *singlet oxygen*, in the process. Singlet* oxygen is an excited state resulting from the pairing of the two unpaired electrons of normal (or triplet state) oxygen. It is thought that singlet oxygen is responsible for the fact the NO is oxidized more rapidly to NO_2. The formation of singlet oxygen is probably enhanced by the presence of unsaturated hydrocarbons, which absorb light and then transfer the energy of their excited states to O_2 in a collision process.

$$:\ddot{O}-\ddot{O}: \quad \text{triplet } O_2 \qquad\qquad :\ddot{O}-\ddot{O}: \quad \text{singlet } O_2$$

$$\text{same spins on electrons} \qquad\qquad \text{opposite spins on electrons}$$

O_3 also reacts with NO to reconvert it back to NO_2.

$$O_3 + NO \longrightarrow NO_2 + O_2$$

This provides a recycling operation when NO_2 reacts with light again.

Atomic oxygen can react with a saturated or an unsaturated hydrocarbon producing free radical species that in turn react with O_2. The net result is the formation of the carbonyl compounds, the aldehydes, and ketones. For example, the unsaturated hydrocarbon 1-butene produces two aldehydes.

$$CH_3-CH_2-CH=CH_2 + O + O_2 \longrightarrow CH_3-CH_2-C\overset{\ddot{O}}{\underset{H}{\diagup}} + \overset{H}{\underset{H}{\diagup}}C=\ddot{O}$$

Note that the carbon chain is broken into two fragments in this reaction.

Another reaction that occurs begins with the formation of a hydrocarbon radical such as the acetyl radical from an aldehyde such as acetaldehyde by loss of a hydrogen atom. The process is also light-initiated.

$$CH_3-\overset{:\ddot{O}:}{\overset{||}{C}}-H \xrightarrow{\text{light}} CH_3-\overset{}{\underset{:\ddot{O}:}{\overset{||}{C}}}\cdot + \cdot H$$

This reacts with O_2 to form the peroxyacetyl radical.

$$CH_3-\overset{}{\underset{O}{\overset{||}{C}}}\cdot + :\ddot{O}-\ddot{O}: \longrightarrow CH_3-\overset{}{\underset{O}{\overset{||}{C}}}-\ddot{O}-\ddot{O}\cdot$$

This new radical reacts with another radical, NO_2 or NO, to produce peroxy-acetylnitrate (PAN), two unpaired electrons forming the new bond.

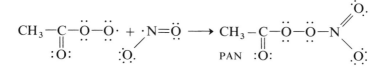

PAN does not have an unpaired electron; hence it has a greater stability than the free radical intermediates. It is the final product of the sequence (Fig. 1.22).

PAN is decidedly an unwholesome compound. It is a major irritant to the eyes and respiratory passages. Furthermore, it has been found that PAN is primarily responsible for the adverse effect on plants and crops that are

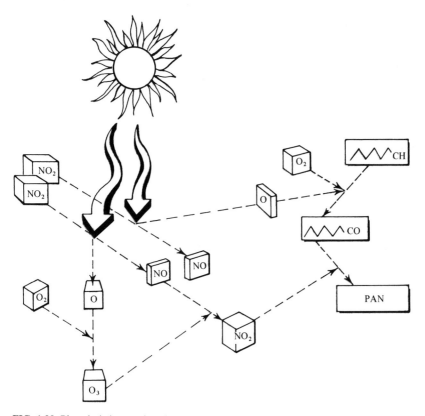

FIG. 1.22 Photolytic interactions between hydrocarbons and nitrogen oxides in the presence of light.

FIG. 1.23 The trees in the foreground show damage believed due to effects from photochemical smog. The photograph was taken in the San Bernardino National Forest, to the east of Los Angeles. (Photo courtesy of U.S. Forest Service.)

definitely observed when smog becomes a problem. In addition, O_3, the aldehydes, and ketones serve as principal irritants in smog.

In retrospect we can see a perverse interaction among the various components arising from interactions of light, natural components in the air, and auto exhaust components. Without the photoactivity of π-bonded compounds we would not observe the formation of singlet oxygen in any amount, and this would slow the formation of NO_2. Without NO_2, the formation of ozone would not take place, and that would slow formation of aldehydes and ketones that form the necessary radicals to start the PAN chains. The absence of any one component would substantially reduce the continuance of the chain and, consequently, smog formation would be reduced.

The Los Angeles Basin offers a perfect setting for smog formation. First, mountains on the north and east prevent the movement of air from the basin sufficiently rapidly to disperse the reactants. Second, everything has been done to encourage the use of the automobile and limit more efficient transportation systems, such as buses or commuter trains, which have a much lower ratio of emission to commuter-miles traveled. Third, there is sunlight nearly the year round. Other locations with better conditions experience smog problems on occasion, however, especially when there is little movement of the air, and it is clear that means must be found to prevent this menace to health and pleasure in our urban centers.

One approach to their limitation is to limit emissions from automobiles. This has been done in recent years, most notably in California. The Environmental Protection Agency has recently proposed federal standards for emissions to be reached by the 1976 model year. The standards relate to emissions of hydrocarbons, nitrogen oxides, and carbon monoxide. It is proposed to reduce the levels of the three pollutant groups by at least 90% from 1970 model vehicles. These levels are being gradually implemented year by year.

The auto manufacturers have suggested that it will be difficult for them to meet these standards and they have advocated further delay because of technological problems, particularly for the nitrogen oxides.

The techniques used to reduce hydrocarbon emissions include exhaust recirculation systems, to oxidize the intermediate products further, and

FIG. 1.24 This view shows a city under a thermal inversion, in which air stagnation has developed because of weather conditions. As a result, the concentration of pollutants has built up to the point at which vision in the city is obscured and health hazards are possible. (Photo by Bill Garlow, *Dayton Journal Herald.*)

TABLE 1.4
UNITED STATES AUTO EMISSION STANDARDS

	Emissions (g/mile)		
	Hydrocarbons	CO	Nitrogen Oxides
Precontrol emission levels	17.0	124.0	5.4
1976 levels anticipated	0.46	4.7	1.0

catalysts in the exhaust system that will complete the oxidation. A major problem is the maintenance of the pollution control system. Among other things, it requires that the individual owner exercise his responsibility for its upkeep.

POWER PLANTS AND AIR POLLUTION

The auto is not the only culprit in air pollution. Two other important causes of air pollution derive less from auto emission than from industrial sources. One is formation of sulfur oxides from the combustion of sulfur. The other is the formation of particulate matter, small particles of carbon and other substances that are emitted.

Sulfur is a serious problem because of its presence in coal. Since coal is used so much in power plants, the oxidation of sulfur to SO_2 and SO_3 and the emission of these noxious gases is a particular problem for the electrical companies and those who live near power stations. Sulfur oxides are anhydrides of H_2SO_3 ($SO_2 + H_2O$) and H_2SO_4 ($SO_3 + H_2O$). They exert acidic effects on some construction materials, particularly stone and masonry, and they are severe irritants and lachrymators.

The problem appears to be solvable by technical approaches. There are two choices. One can remove the sulfur from the coal or one can remove the sulfur oxides from the stack gases. The latter is more feasible in practice. Several possibilities for reactions that will remove SO_2 are under study. In one, SO_2 is reacted with limestone producing less harmful CO_2.

$$CaCO_3 + SO_2 \longrightarrow CaSO_3 + CO_2$$
limestone

Another involves oxidation of SO_2 to SO_3 using vanadium pentoxide, V_2O_5, as the catalyst. SO_3 is the acid anhydride of sulfuric acid, H_2SO_4, so that by dissolving the SO_3 in water, one can produce sulfuric acid, which is a product of considerable industrial importance, and as a by-product one could reduce the expenses of sulfur oxides control.

Yet another method under study neutralizes the acidic SO_2 by dissolving it in water that is slightly basic (pH 8.1). The reaction with hydroxide ion converts the gas to a soluble anion, the sulfite anion.

$$SO_2 + 2OH^- \longrightarrow SO_3^{2-} + H_2O$$

Particulates may be removed from gas emission by a process known as *electrostatic precipitation*. This technique is based on the fact that there are some unneutralized charges at the surface of particles, and in presence of high voltage these particles can be precipitated onto the surface of the precipitator, which is acting as a giant electrode.

Both particulate and sulfur oxide problems can thus be solved by technical changes in how industry treats gas emissions. This will require expenditure by industry or by government in support of industry. In either event, all of us as consumers or as citizens will bear the cost. Again the problem is whether we are willing to pay the necessary price for cleaner air and better health.

SUGGESTED READING

1. M. King Hubbert, "Energy Resources" in *Resources and Man*, A Study and Recommendation by the Committee on Resources and Man of the Division of Earth Sciences, National Academy of Sciences and National Research Council. W. H. Freeman and Co., San Francisco, Calif., 1969, pp. 157–242. This authoritative study examines energy sources and prospects for the future. It discusses the problems of predicting when the likely depletion of existing and yet-to-be-discovered resources of fossil fuels will occur.

2. *Cleaning our Environment—The Chemical Basis for Action*, American Chemical Society, Special Issue Sales, 1155 Sixteenth St., N.W., Washington, D.C. 20036. This work discusses chemical aspects of the air environment and problems associated with both industrial and automotive sources of pollution.

3. *Vanishing Air* by John C. Esposito, Grossman Publishers, New York, N.Y., 1970. This study examines air pollution and its sources. It is very critical of abatement efforts of the National Air Pollution Control Agency (NAPCA), now a part of the Environmental Protection Agency.

4. *Man's Impact on the Global Environment*, a report of the Study of Critical Environmental Problems (SCEP), The MIT Press, Cambridge, Mass., and London, England. This study deals with global effects of pollutants, including CO_2 and particulates.

5. "Implications of Rising Carbon Dioxide Content of the Atmosphere" in K. E. Maxwell, *Chemicals and Life*, Dickenson Publishing Company, Inc., San Francisco, Calif., 1970. This discusses the implication on the earth of a rising atmospheric CO_2 level in more detail than treated here.

6. D. O. Sullivan, "Air Pollution," *Chemical and Engineering News*, **48**(24), 38–58 (1970). A description of air pollution problems relating particularly to industrial and automotive problems.

7. "Energy for the Future," National Academy of Sciences Symposium, *Proceedings of the National Academy of Sciences of the U.S.A.*, **68**(8), 1920–1943 (1971). This symposium considers many aspects of the energy problem, in particular, relating to needs for future development of nuclear reactor technology.

8. C. Starr, "Energy and Power," *Scientific American*, **225**(3), 37–49 (1971); M. King Hubbert, "The Energy Resources of the Earth," *ibid.* 61–70; David M. Grates, "The Flow of Energy in the Biosphere," *ibid.*, 89–100. These articles, appearing in September, 1971, are concerned with many aspects of the energy problem described in this section.

9. "The Drive to Control Auto Emissions," *Environmental Science and Technology*, **5**(6), 492–495 (1971). This article describes progress toward EPA standards for emissions for 1975 model autos.

10. Dean E. Abrahamson, *Environmental Cost of Electric Power*, Scientists' Institute for Public Information Workbook, SIPI, 1970. This article is a thorough and thoughtful analysis of all the environmental costs of air pollution.

11. *Air Quality Criteria Statements.* In recent years the Environmental Protection Agency has published Air Quality Criteria Statements on certain air pollutants that are designed to relate the state of knowledge concerning dangers associated with their emission. The statements are designed to serve as guidelines in limiting air pollution. Among statements issued to date are "Photochemical Oxidants" (NAPCA Publication AP-63, March 1970); "Carbon Monoxide" (AP-62), March 1970; "Hydrocarbons" (AP-64), March 1970; "Particulate Matter" (AP-49), January 1969. These may be obtained from Superintendent of Documents, U.S. Government Printing Office, Washington, D.C. 20402.

2

Pollution of Our Natural Waters by Nutrients

> Everything was fine until the last scene of the opera [Aida],
> during which the hero and heroine were placed in
> an airtight chamber to suffocate. As the doomed pair
> filled their lungs, Eliot called out to them, "You will last
> a lot longer, if you don't try to sing." Eliot stood, leaned far
> out of his box, told the singers, "Maybe you don't know
> anything about oxygen, but I do. Believe me, you must not sing."
>
> *Kurt Vonnegut, Jr.*
> God Bless You, Mr. Rosewater
>
> © *Seymour Lawrence/Delacorte Press*
> *Dell Publishing Co.*

Chemical Orientation The addition of materials that serve as nutrients in the biosphere will be the topic of this chapter. These nutrients come from fertilizers, human and animal wastes, industrial operations, and detergents. A very important chemical principle, that of chemical equilibrium, is involved numerous times in this chapter. You should review chapters in your chemistry text on equilibrium involving acid-base reactions and the formation of precipitates and complexes from ions in solution. Concepts of Lewis acids and bases and the nature of the action of chelating agents in forming complexes in solution should be reviewed. The concepts of oxidation and reduction should be reviewed, as well as some aspects of the chemistry of nitrogen and phosphorus.

INTRODUCTION

There are two kinds of pollution of an aquatic environment, be it a river, a lake, or the sea. DDT is a good example of one type of pollutant, occurring when technological advance produces a new material that is not known in nature; that is, it is *synthetic* in origin. If such a substance is also chemically stable in the environment, the situation is further complicated. Large quantities of such a substance produced, dispersed, used, and then discarded by man may persist and the amounts of it present will increase slowly. Chances are then good that it will interact adversely with the environment, because organisms and ecosystems have had no experience in dealing with the compound. Problems associated with DDT and other persistent compounds will be treated in a later section.

The other pollutant situation involves the introduction of large quantities of a substance which is found in nature and can be accommodated but which is present in much greater abundance than the ecosystem has had to cope with normally. In other words, the *concentration* of the substance is too high. Indeed the substance may be a necessary one for the health of plants and animals. Sometimes this is a worse situation because it can disrupt the biological equilibrium, the interactions of populations of different plants and animals with one another. This kind of pollution is principally involved in what has happened to Lake Erie and many other bodies of water. The ecological disruption found there is due to the introduction of large quantities of nutrient materials for plants. Of course these are necessary for the growth and health of the aquatic biological community, but in high concentrations they cause a runaway cancer-like overgrowth leading in the extreme to collapse of the ecosystem. Both inorganic and organic compounds are involved in this way. In this section we look at both kinds of nutrients, at the sources, and their possible control.

THE ROLE OF NITROGEN

Nitrogen, atomic number 7, like its periodic table neighbor carbon at 6, is intimately involved in the chemistry of living systems. Like carbon, its role tends to link the fates of plant and animal life together and to their habitat. Also like carbon, the involvement of nitrogen in biology is due to reactions that change the oxidation state of the atom by the addition or removal of electrons.

Some Chemical Comments

We list in Table 2.1 the most important simple species of nitrogen together with their oxidation states. The oxidation state is determined by assigning to all oxygens in the species a state of -2; to all hydrogens $+1$, and to N, an oxidation state that will balance the overall charge of the molecule or ion. Any change that removes electrons will increase the oxidation state of the nitrogen, moving to compounds of higher oxidation state. A reduction* process, adding electrons, will move the nitrogen down the scale. Other events in the reaction are required to change the numbers of oxygen and hydrogen atoms present—attention here focuses only on the net change in oxidation of nitrogen, as reflected in the increasing or decreasing oxidation state.

The normal introduction of nitrogen into an undisturbed aquatic ecosystem from outside happens in three ways.

1. Nitrate ion is dissolved from rocks or soil and taken up by organisms (Table 2.1).
2. N_2 from the air is "fixed" by biochemical reactions of reduction leading to ammonia or organic derivatives of ammonia.
3. Ammonia, present in small amounts in air and water, is absorbed and incorporated into organic molecules.

TABLE 2.1
OXIDATION STATES OF NITROGEN

Name	Structure	Oxidation State of N	
Nitrate ion		-2×3	Oxygen charge
		$+5$	Nitrogen charge
		-1	Charge on ion
Nitrite ion		$+3$	
		$(-2 \times 2) + 3 = -1$	
Nitrogen	$:N{\equiv}N:$	0	
Hydroxylamine		-1	
Diimide (unstable)	$H-\ddot{N}{=}\ddot{N}-H$	-1 (each N atom)	
Hydrazine		-2 (each N atom)	
Ammonia		-3	

Once absorbed, the nitrogen is interconverted by biochemical oxidation-reduction processes that are usually catalyzed by enzymes. Also it will react with organic materials producing important organic nitrogen compounds such as amino acids, plant alkaloids, and nucleic acids.

As an example of this, consider the reaction of ammonia with α-ketoglutaric acid, which is formed by carbohydrate decomposition.

$$HO_2C-\underset{\underset{:O:}{\|}}{C}-CH_2-CH_2CO_2H +:NH_3 + 2H^+$$

$$\longrightarrow HO_2C-\underset{\underset{:NH_2}{|}}{CH}-CH_2-CH_2-CO_2H + H_2O$$

The reaction produces the amino acid called *glutamic acid* by inserting an amino group at a position formerly occupied by a carbonyl group ($>C=O$). Thus ammonia is incorporated into an organic compound.

The conversion of nitrate ion to lower oxidation states is carried out in plants through a series of enzyme-catalyzed reductions. This is the most important process in the nitrogen cycle because it also ties the inorganic and organic environments together. The incorporation by the conversion of gaseous N_2 to NH_3 is accomplished by bacteria centered in the root of leguminous plants, such as alfalfa or bean plants. In aquatic systems some species of blue-green algae accomplish the same thing. This reduction process may be written

$$2N_2 + 6H_2O \longrightarrow 4NH_3 + 3O_2$$

It proceeds through several intermediate stages before reaching the final one. This equation only summarizes the overall reaction.

The oxidation process of nitrification, the conversion of NH_3 to nitrate by bacteria, also is important.

Figure 2.1 shows details of the nitrogen cycle involving atmosphere, living matter, decomposing bacteria, and the soil.

PHOSPHORUS COMPOUNDS AS NUTRIENTS

Phosphorus in plants occurs only as derivatives of phosphoric acid, H_3PO_4. The oxidation state of phosphorus is $+5$. Successive loss of one or more protons leads to the respective anions dihydrogen phosphate, $H_2PO_4^-$; hydrogen phosphate, HPO_4^{2-}; and phosphate, PO_4^{3-}. When the protons

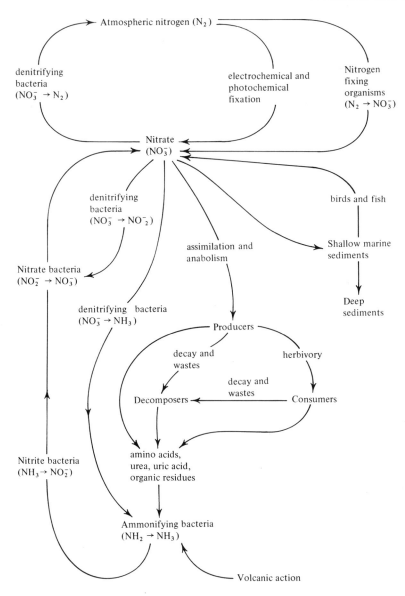

FIG. 2.1 The nitrogen cycle. (From Edward J. Kormondy, *Concepts in Ecology,* Prentice-Hall, Inc., Englewood Cliffs, N.J., 1969, p. 44.)

are replaced with organic groups, phosphate esters form. An example of this is α-D-glucose-6-phosphate, an ester formed between a hydroxyl group of a simple sugar (see Fig. 1.3), and a phosphate group. In forming the ester a molecule of water is also produced.

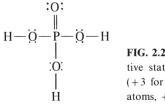

FIG. 2.2 Phosphoric acid: Oxidative state of P is +5 as shown. (+3 for 3 H atoms, −8 for 4 O atoms, +5 for phosphorus.)

Such phosphate esters are important intermediates in carbohydrate metabolism.

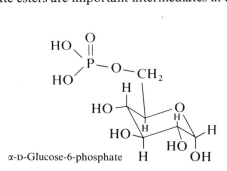

α-D-Glucose-6-phosphate

Phosphate groups can form esters with themselves by combinations of two or more phosphate groups. The structures of ATP and ADP are a good example (see p. 22). Note that there are two phosphates linked together in ADP; three, in ATP.

The nutritional role of phosphates is crucial to both plants and animals. Phosphates are required in the metabolism of carbohydrates and in the energy-releasing reaction of ATP. They are present in nucleic acids and in proteins present in the cell nucleus associated with the nucleic acids. Phosphate is essential to the development of bone tissue.

An insufficiency of nitrogen or phosphorus can limit growth of an individual or a community of plants. The external application of nitrates (or ammonia) and phosphates will enhance growth and produce sturdier plants and higher yields of crops. Thus, these elements have played central roles in agriculture as fertilizer and are in large measure responsible for increased yields of agricultural crops in the United States and other developed agricultural areas. The same role of stimulation of plant development in an aquatic environment can be detrimental when too much of these inorganic nutrients are present.

EUTROPHICATION: ECOLOGICAL OBESITY

Let us summarize the nutrient requirements of plants. Since animals feed on plants, either directly or indirectly, this becomes a listing of nutrient requirements for both:

1. Light is required for photosynthesis.
2. Water and CO_2 are needed as the reactants in the photosynthetic process.
3. Phosphate must be available.
4. Nitrogen, usually as nitrate ion, must be available.
5. Small amounts of many other minerals are required.

One might think that plant growth and vitality would be enhanced by increasing any of the needed nutrient factors. The situation is a little more complex than that. Just as the health and growth of a person will be inhibited by the absence of a single necessary component, like vitamin C, the healthy development of a plant community will be limited by the essential nutrient that is present in the smallest amount compared with the amount needed. The others will thus be present in some excess. The excess will not usually be an adverse factor in the health of the ecosystem. By turning the argument around, we reach another conclusion. If a given nutrient is not the limiting one, addition of more of it will not disrupt the ecosystem.

On the other hand, if the concentration of this *limiting nutrient* is increased because of the activity of man or by some other change, then the plant population will increase. Aquatic plant overgrowth occurs because of such nutrient enrichment. The complex equilibrium interaction of the plant and animal life of the stream is disrupted. In the normal situation in a lake, ecologists view the equilibrium to include three living components. The *producers* are the plants, principally algae, which take in CO_2 and produce carbohydrates. The *consumers* are herbivores, which feed on these plants, regenerating CO_2 again through respiration. Carnivores are secondary consumers since their nutrients are ultimately linked to the plants through a food chain, with several intermediate links possible in the chain. The *decomposers* interact with dead plant and animal tissues and with wastes from living plants and animals. They convert plant and animal matter back into inorganic components, CO_2, nitrate and phosphate ions, and so on. These are available for recycling by producers and consumers. In the process the bacteria consume O_2 and produce CO_2, as do the consumers. Thus we see a cycle of the inorganic nutrients being processed through the biological community. Energy enters from outside in the photosynthesis reaction and so do dissolved inorganic nutrients.

Man's activity can disrupt the natural equilibrium. Suppose we take as a model for study a mountain lake in the western United States, a pristine isolated gem. Because of its isolation the lake has been used by only the few fishermen who were willing to take the time and trouble to make their way over difficult terrain to reach it. Those few valued the lake not only for its abundant gamefish but for its isolation. Then change comes. A freeway is constructed near the lake and a developer sees the potential for recreation arising from this new accessibility. A resort is constructed, and large numbers of people come to enjoy the lake and the surrounding area. Other resorts follow. That in itself does not constitute a severe ecological disruption, but there is the problem of the wastes produced by these people.

Unfortunately sewage, food wastes from the restaurants, and other organic matter is produced from all this human activity. Because it is cheapest and easiest and because no harm is thought to come by it, the wastes are dumped into the lake. The quantity of organic carbon made available to the decomposers in the lake is markedly increased by this action. So too are the inorganic nutrients that they require. As they metabolize the organic compounds. phosphates and nitrates are produced as products, increasing further the concentrations of these nutrients. The decomposers also consume O_2 and produce CO_2 thus altering the concentrations of these gases in the waters of the lake. Moreover the population of the decomposers increases due to their favorable food supply situation. This condition is reversed by the response of the producers to the increased supply of CO_2 and inorganic nutrients. Their population rapidly increases and O_2 is resupplied. In the normal case the result is passed along the food chain by stimulating the consumer population. Completing the cycle, the decomposer population is increased again by the increased supply of plant and animal matter available to it for decomposition. We find ourselves on a slow upward spiral of increasing vitality in the lake.

In the extreme case, however, the development of plants becomes very extensive. An *algae bloom* gives a green cast to the water and a film of algae appears on the surface. Sooner or later an ecological catastrophe strikes the lake. It happens when the requirements of decomposition of the plants on the oxygen supply outstrips the O_2-producing capacity of the plant producers. The oxygen supply becomes too low, and the population of consumers is seriously cut or even decimated because of their dependence on O_2. Only anaerobic bacteria, which do not require O_2, and the algae survive. Plant decomposition slows down, and the presence of large quantities of incompletely decomposed plant and waste matter imparts a putrid smell to the lake. The lake becomes more like a swamp than a body of water.

Under normal conditions where the input of nutrients is only slightly faster than outflow, this process, called *eutrophication*, will take place very slowly. Under the stimulus of large quantities of externally added nutrients the process accelerates.

Lake Erie has been cited as an important instance of eutrophication on a scale far larger than that which we have just depicted, although eutrophication has not reached advanced stages. There has clearly been large increases in the rate of eutrophication in the last century. In the last 50 years, limnologists* believe that the addition of nutrients has caused the aging

FIG. 2.3 The photo at top, taken at Lake Tahoe on the California–Nevada border, indicates the kind of overgrowth of algae resulting from excess of nutrients in the water. (Photo courtesy of EPA-DOCUMERICA—Belinda Rain.) The photo at bottom shows the kind of effluents resulting from industrial, municipal, or agricultural sources which can lead to the excess nutrients described. (Photo by Walt Kleine, *Dayton Journal Herald*.)

TABLE 2.2
THE INCREASE IN NITROGEN AND PHOSPHORUS IN
LAKE ERIE BETWEEN 1942 AND 1965–66[1] (in μg/l)

	1942	1965–66
Available N as NH_3 and NO_3^-	261	330
Soluble phosphorus (μg P/1)	7.5	36

[1] *Agriculture and the Quality of Our Environment*, AAAS Publication 85,
The Plimpton Press, Norwood, Mass., 1967.

process in Lake Erie to advance to a point that would not have occurred
naturally for 20,000 years. Similar effects of varying severity have also been
noted in Lake Baikal and the Caspian Sea in the Soviet Union, in the Sea of
Galilee in Israel, and at Lake Tahoe in California and Nevada.

Research has established that the concentrations of nitrogen and phosphorus nutrients has increased in recent decades, as seen in Table 2.2.

Which nutrients are to blame? Is it nitrate ion, ammonia, and other
nitrogen compounds? Is it a trace mineral or ion? Most ecologists believe
that phosphate is intimately involved in the Lake Erie situation. The data
for Lake Erie implicates phosphates since its concentration has increased
fivefold, while N has increased less sharply. Some believe that the limiting
nutrient is nitrogen or that the ratio of both phosphate and nitrogen is critical.
It is quite possible that the differences between bodies of water, and the
nature of the neighboring soil and the water feeding them, will make the
limiting nutrient different.

Approximately 150,000 lb of phosphates enter Lake Erie daily. The
sources are listed in Table 2.3.

TABLE 2.3
PHOSPHATE SOURCES IN LAKE ERIE[1]

Source	Daily Quantity (1b)
Runoff from rural lands	20,000
Detergents	70,000
Entering from Lake Huron	20,000
Human excreta	30,000
Runoff from urban land	6,000
Industrial discharges	6,000

[1] K. Sperry, "The Battle of Lake Erie; Eutrophication
and Political Argumentation," *Science*, **151**, 351 (1967).

TABLE 2.4
SOURCES OF NITROGEN ENTERING LAKE ERIE[1]

Source	Annual Quantity (millions of pounds)
From soil organic matter	110
Livestock manure	56.5
Fixation by soil organisms	56.5
Added in rainfall	27
Fertilizer	39
Human wastes	12

[1] Hearings before the Subcommittee on Air and Water Pollution, Committee on Public Works, United States Senate; ninety-first Congress, 1970, p. 770.

The introduction of nitrogen as nitrate, ammonia, or organic nitrogen compounds is estimated to arise from the areas shown in Table 2.4. About 300 million lb of nitrogen are introduced per year by all these routes. Although the nitrogen quantity is much greater than phosphate, the reader should keep in mind that the limiting nutrient is still the important one, and much less phosphate than nitrogen is required for plant growth.

Apart from doubts about the specific cause of eutrophication in Lake Erie and similar bodies of water and the need to understand what general and local factors cause this form of pollution, it is quite clear that a number of things can be done even before the problem is completely understood. Every indication is that when the nutrients are removed, much can be accomplished to restore a lake or stream to its former condition.

In the next section we examine various human activities that contribute nutrients to our waterways. We shall try to assess the effect of each, how the nutrient could be removed if that is necessary, and at what cost.

THE SOLUTIONS TO THE EUTROPHICATION PROBLEM: GETTING AT THE SOURCES

The added nutrients are due mainly to technological advancement and the tendency to concentrate both human beings and animals into small areas. Before discussing these sources in detail we enumerate them:

1. *Agricultural fertilizers* release large quantities of inorganic nitrogen and phosphates into the environment.
2. *Laundry detergents* represent a major source of phosphates.
3. *Human wastes* resulting from output of untreated or incompletely treated sewage are responsible for organic carbon compounds in particular. They also are rich in nitrogen and phosphorus compounds.

4. *Animal wastes* result from concentration of livestock or poultry in large feed lots due to modern feeding practices. They can cause serious local pollution problems because of the resultant high concentrations.
5. *Industrial pollution* involving nutrient materials is of particular concern in the paper- and food-processing industries. Most chemical-processing industries also produce oxidizable organic compounds in their effluents. The nutrients problem caused by industrial activity will be discussed in Chapter 3, "The Organic Chemicals Industries."

Phosphate from human wastes and from detergents will tend to be concentrated in the sewage of municipalities. The fact that sewage is dumped into a lake or stream in a single place by an urban sewage disposal plant means that high concentrations of the nutrients will be present at that site. Chemists and biologists have always felt that this problem would be alleviated by dilution as the nutrients were dispersed through the whole body of water. We have recently found that the mixing process is not as simple as theoretical models suggest. The nutrients tend to persist, incorporating themselves into the biomass* near the point of emission. This is a more serious problem in a lake than in a flowing stream where effluents are dispersed somewhat more rapidly.

It should also be borne in mind that the minerals of the lake or stream bed also can be a source of phosphates and nitrates. The present arguments involving limnologists, detergent manufacturers, fertilizer manufacturers, sanitary engineers, and others relate to the question of which sources of phosphates are to be blamed, if phosphates are responsible at all. The opinions generally held by various interests tend to place blame on someone else or some other industry. Unfortunately, the data are not clear-cut or consistent from one body of water or situation to another.

Laundry detergents add about 2 billion lb of phosphates to the effluents of urban centers. Estimates are that this source accounts for about two-thirds of the total phosphorus found in municipal wastes. The rest comes from human waste and from industrial effluents.

While they produce nitrogen and phosphorus, the principal component provided by human wastes is the supply of carbon compounds capable of further oxidation found in the feces. The extent to which this is a problem depends on how thoroughly sewage is treated. This will be discussed shortly.

Animal wastes present essentially the same problem as that for human wastes. The domestic animal population of the United States produces the waste equivalent of 2 billion humans. If this is dispersed over wide areas, there is no problem, but the recent tendency has been to concentrate livestock and poultry in densely populated feedlots. There are 34 million beef cattle and calves slaughtered annually in the United States. Of these, 23 million are fattened in feedlots, and half of these are in herds of more than

FIG. 2.4 This flock of turkeys is a typical example of the problem that exists in large aggregations of livestock or poultry. The waste materials produced by these animals can cause a serious removal problem for the farmer and can present an environmental hazard if the wastes reach a lake or stream. (Photo courtesy of *Dayton Journal Herald*.)

1,000 head. The rest are fattened by range or pasture feeding. Of the 12.5 million dairy cattle, more are confined in larger herds than in earlier times. There are about 3 billion domesticated poultry, 67 million swine, and 21 million sheep. Chicken flocks as large as 1 million birds are sometimes found, and such an assemblage produces as much solid wastes as a city of moderate size. There have been many instances of nutrient pollution from such large congregations of livestock. A prominent example is offshore pollution in Long Island Sound from large flocks of ducks.

In the following sections we shall discuss in further detail some of the potential nutrient sources and means of their control.

THE CHEMISTRY AND ECOLOGY OF INORGANIC FERTILIZERS

The same substances that enrich a lake or stream and thus cause pollution by plant overgrowth are necessary to the survival of man under present circumstances, and for the same reason, for, as our population grows, our land resources remain unchanged. They even decrease somewhat because of urbanization. Thus production must increase per unit area. It becomes of interest to use means of stimulating the growth and vitality of plants by providing nutrients. Natural fertilizers, the wastes of plants and animals, provide most of the required nutrients, but this source is supplemented by the use of inorganic salts. These fertilizers have three elemental components that are important in plant growth. Nitrogen is applied as solid nitrate salts or as ammonia under pressure in the liquid state or in aqueous solution. Phosphorus is added as phosphate minerals. The element potassium is added as the mineral potash, which is a mixture of various potassium salts including the chloride, nitrate, and sulfate. The relative composition of nutrients is varied depending on crop and soil needs. About 25 million tons of fertilizer are used in the United States each year.

To make fertilizer phosphates, tricalcium phosphate, $Ca_3(PO_4)_2$, a mineral form of phosphate, may be converted to calcium dihydrogen phosphate by treatment with sulfuric acid.

$$Ca_3(PO_4)_2 + 2H_2SO_4 + 4H_2O \longrightarrow Ca(H_2PO_4)_2 + 2CaSO_4 \cdot 2H_2O$$

The reason for this reaction is that calcium dihydrogen phosphate is soluble in water and will dissolve into the soil and be absorbed after application. This important form of phosphate is called *superphosphate*.

The reaction of ammonia with phosphoric acid produces the soluble ammonium phosphate and of course provides both phosphorus and nitrogen nutrients. Monoammonium or diammonium phosphate will be produced depending on the proportions of NH_3 and H_3PO_4 used.

$$NH_3 + H_3PO_4 \longrightarrow NH_4H_2PO_4 \quad \text{ammonium dihydrogen phosphate}$$

$$2NH_3 + H_3PO_4 \longrightarrow (NH_4)_2HPO_4 \quad \text{diammonium hydrogen phosphate}$$

Note that this process is a simple proton-transfer reaction between an acid H_3PO_4 and a base NH_3, the proton moving from the proton-donating acid to the proton-receiving base.

$$NH_3 \longrightarrow NH_4^+$$
base   conjugate acid

$$H_3PO_4 \longrightarrow H_2PO_4^-$$
acid   conjugate base

The production of ammonia principally occurs through a famous and enormously important chemical reaction known as the Haber process. Its inventor, Fritz Haber, was a German chemist of the early twentieth century.

The process involves the preparation of ammonia from its elements, N_2 and H_2. The reaction is

$$N_2 (g) + 3H_2 (g) \rightleftharpoons 2NH_3 (g)$$

The reaction is in equilibrium, which means that a dynamic balance exists between reactants and products which interconvert so that the concentrations of each reactant and product remain unchanged. In fact the reaction is not favorable to the production of ammonia because the reaction is slow and because the equilibrium concentrations lie on the side of N_2 and H_2 rather than NH_3.

To compensate for this, several steps are taken to make the process feasible (Fig. 2.5). The reaction is catalyzed by a solid catalyst of iron and ferric oxide, which accelerates the rate at which the equilibrium is attained. Then the equilibrium is subjected to high pressures, greater than 600 atm. The influence of increased pressure is to shift a reaction such as this one in the direction yielding fewer moles of gases, which means toward the ammonia

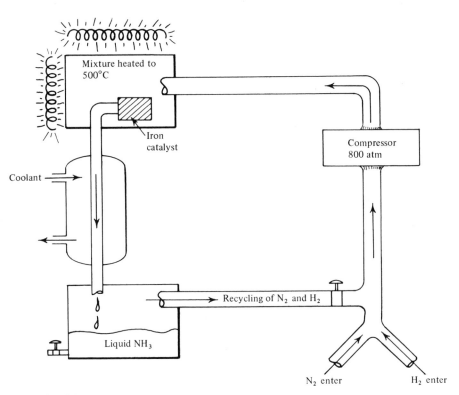

FIGURE 2.5

side. The third step that is done is to remove ammonia by condensing it. Since its boiling point is much lower than either nitrogen or hydrogen, it alone will condense when the equilibrium mixture is cooled. The removal of NH_3 from the gas phase will serve to shift the reaction in the direction of the production of more ammonia.

By this process nearly 12 million tons of synthetic anhydrous ammonia is produced annually in the United States. Indeed the synthetic fixation of atmospheric nitrogen represented by the Haber process rivals the activity of leguminous plants, in converting nitrogen in the air into a form that can be utilized as a nutrient for plants and animals. Most of the ammonia produced goes, directly or indirectly, into fertilizers.

The large quantities of fertilizers used could represent a serious nutrient water pollution problem, but only if the fertilizers are washed out of the land and into the streams by rainfall. It has been shown that phosphates are absorbed into soil particles and made insoluble in large measure after application, and while there is some phosphate runoff, the quantities from this source are less than from others into most lakes and streams. The situation with nitrogen is different since nitrate salts are water-soluble. Some nitrate runoff is possible, although it has not been established how serious this problem is as far as eutrophication is concerned.

THE STRANGE STORY OF METHEMOGLOBINEMIA

The evidence suggests, as we have seen, that most phosphate fertilizer applied to the soil is retained, so that it does not appear to be the major cause of eutrophication. On the other hand, nitrate salts are soluble. This means that some nitrates can be carried off on the surface in rainwater or deeper into the soil by leaching. This possibility is responsible for a new health problem in parts of central California. As we shall see, a strange set of uncanny circumstances is responsible.

The Central Valley of California is one of the most important agricultural areas of the United States, producing enormous quantities of fruits and vegetables. With rich volcanic soil and a long growing season, only sufficient rainfall is lacking; irrigation has long been practiced in the area, particularly in the drier southern area drained by the San Joaquin river. For years irrigation was dependent on underground supplies but as more and more irrigation went on, it became clear that alternate water supplies were needed to bring water into the valley from the snow-rich Sierra Nevada Mountains on the east. As water was being taken out faster than it was replaced, the water table had dropped by as much as 170 ft by 1950. The opening of the Friant-Kern Canal provided the outside water source that irrigation required.

As irrigation had proceeded through the years, the use of fertilizers had increased in the San Joaquin Valley as it had elsewhere in American agriculture, and some leaching of nitrates deep into the soil had taken place. When new water sources were introduced, the water table began to rise again because the underground sources were not being used. As the levels increased, they came into contact with the nitrates newly present in the soil. Because the nitrates are soluble, they dissolved in the groundwater and their concentration in water thus increased. While the groundwater is no longer used for irrigation, it continues to serve as the source of drinking water for the farms and communities in the valley.

There is nothing toxic about nitrates per se, and they are present in all body fluids. If nitrates should be reduced to nitrites (see p. 67) however, the situation changes, because nitrite can be mildly toxic in the bloodstream by causing a change in hemoglobin, the molecule that transports oxygen in the blood (p. 174). The key structural feature of hemoglobin responsible is the presence of iron ions in the +2 ferrous oxidation state. If nitrite ion is present in the bloodstream, it can effect the oxidation of hemoglobin to methemoglobin, the same molecule, except that the ferrous ions have been oxidized from the +2 to the +3 oxidation state. Nitrite ion is reduced at

Hemoglobin + nitrite $\longrightarrow$ methemoglobin + nitrogen reduction products

iron in +2   iron in +3
oxidation state   oxidation state

the same time. Methemoglobin differs from hemoglobin only in that it cannot transport oxygen because of the change in oxidation state. Thus the result is the same as carbon monoxide poisoning. If enough hemoglobin is removed, cell function begins to be impaired by oxygen insufficiency leading in the extreme instance to death.

Nitrites can only be present if they are reduced from nitrates and this does not normally happen in the human digestive tract. Some bacteria are present in the lower bowel that are capable of reducing nitrate to nitrite but they cannot live in the stomach or small intestine because the acidity is too high there. This nitrate converts to nitrite only below the point where major absorption into the bloodstream takes place.

Thus nitrates are not toxic to humans, with a single very important exception. It happens that the pH* in the stomach of infants is higher; thus the acidity is lower than for adults or older children. For this reason the bacteria can invade the upper reaches of the digestive tract, reduce the nitrate to nitrite, and enable its consequent absorption into the bloodstream where it can lead to the disease methemoglobinemia. Thus water given in the infant's formula can prove toxic to him if it has a high concentration of nitrate salts.

The upper level of nitrate concentrations in drinking water that will not lead to methemoglobinemia has been set at 45 mg/l (milligrams per liter).

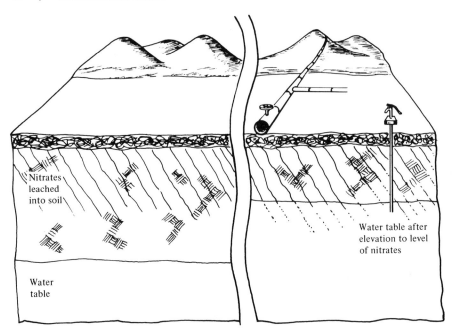

FIG. 2.6 The left diagram depicts the lowering of the water table due to use in irrigation and the leaching of nitrates from fertilizers and other sources by irrigation water. The right diagram shows the present situation: irrigation water has been brought in from outside, causing the water table to rise to the level at which nitrate salts are present and dissolving them so that some nitrate is contained in water used for drinking.

In several communities, drinking water exceeds this, ranging to 80 or 100 mg/l. This has led to health warnings in some instances and a few known cases of the disease. Similar levels have been found in other parts of the world, most notably in Israel, where the same conditions and events occurred.

The situation required three events: the lowering of the water table, the addition of nitrates to the land, and then the subsequent rise again in ground-water levels. The story illustrates the complexities of ecology and how man can unwittingly cause a problem while operating with all good intent. As we have seen and shall see, this is a recurring theme in the environmental crisis.

SOAPS AND DETERGENTS: THE SCHIZOPHRENIC MOLECULES

Soaps and detergents are rather unique molecules. The uniqueness arises from special structural properties that they possess. In order to understand both the action of detergents and their environmental impact, we need to look first at their structures.

Some Chemical Comments

Substances tend to be divisible into categories of ionic and covalent compounds. An ionic salt such as potassium chloride is composed of ions,* or charged atoms. Ions are formed by the gain or loss of electrons from neutral atoms. If an electron of a potassium atom is transferred to a chlorine, K now has a positive charge and Cl has a negative charge. The force holding them together is an electrostatic attraction between the positive and negative charges.

On the other hand, the covalent* compounds form bonds between atoms by sharing electrons. Instead of occupying orbitals located in just one atom, the electrons occupy molecular orbitals spread through a large portion of the molecular volume. The fact that an electron in the molecular orbital is attracted by two or more nuclei instead of just one is the factor that makes the molecule more stable than it would be if the bond were broken.

Real molecules may have properties somewhere between the two extremes of complete electron transfer of an ionic compound and equal sharing of an electron pair as in the covalent case. Thus, one finds covalent bonds in which the electron pair is attracted more strongly to

FIG. 2.7 Three examples illustrate differences in bonding. In Cl_2, the two electrons forming a bond between the chlorine atoms are equally distributed around and about the two nuclei. In KCl, the two electrons are transferred completely to the chlorine atom, giving each atom a complete ionic charge. Indeed KCl forms a complex lattice* in which each ion interacts with several ions of the opposite charge in a three-dimensional crystalline lattice. HCl, a polar* molecule, represents the intermediate situation. Because the electronegativities are different, the electrons are attracted more toward the chlorine. This imbalance creates a partial positive charge at the hydrogen end of the bond and a partial negative charge at the chlorine end.

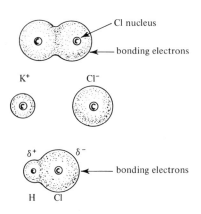

one nucleus than another. HCl is an example. The electrons are more attracted to the chlorine because of its higher electronegativity. The electrons are unevenly dispersed over the molecule and a partial charge develops. The hydrogen end of the molecule is slightly positive; the chlorine end is slightly negative. In fact, completely equal sharing would be expected only when the two atoms were the same, in a symmetric molecule like H_2. In Fig. 2.7 are shown these three types of differing polarity in bonding.

The properties of a substance are related to the degree of polarity of the molecules. Take the case of solubility. Ionic and polar molecules are often soluble in water, which is itself polar, because electrical attraction between water and solute molecules or ions is possible.

Water polarity is caused by the fact that oxygen exerts a stronger attraction on the electron pair in the oxygen–hydrogen bond than the hydrogen nucleus does. This effect is demonstrated for a polar molecule A—B that is interacting with two polar water molecules. Each end of the dipole is attracted to the opposite charges in water molecules.

(δ means *partial*.) There is not a full positive charge as one would find for a cation or anion.

Nonpolar or weakly polar molecules are not soluble in water because the stabilizing attractions between water molecules that are disrupted when the solute is present are not compensated for by the attractions between solute and water molecules. These compounds are soluble in weakly polar or nonpolar solvents such as carbon tetrachloride. Hydrocarbons such as found in gasoline and many other organic compounds are in this solubility category.

Soaps and detergents are molecules that are *both* polar and nonpolar. As a result they have a very special kind of solubility, and a very useful one. They find themselves as molecules with split personalities.

The most common soap, used since antiquity, is sodium stearate. Its preparation is by the reaction of tallow or other animal fats with a source of alkali. Tallow has a primary component, triglyceryl stearate, that has the

structure

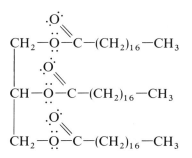

This is a large molecule but it is not a complicated one. It is always helpful in looking at a complex structure to try to recognize fragments. One notes first that there are three long carbon chains. Each is linked to the same 3-carbon unit through *ester* functional groups. Esters are derivatives of carboxylic acids and alcohols, as described in the equation

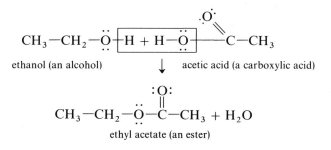

ethanol (an alcohol)   acetic acid (a carboxylic acid)

$$CH_3-CH_2-\overset{..}{\underset{..}{O}}-\overset{:O:}{\overset{\|}{C}}-CH_3 + H_2O$$

ethyl acetate (an ester)

In the case of triglyceryl stearate the components are an 18-carbon acid, stearic acid, and an alcohol, glycerin, which has three OH groups on the same molecule.

$$\begin{array}{l} CH_2-\overset{..}{\underset{..}{O}}-H \\ CH-\overset{..}{\underset{..}{O}}-H \quad + \quad 3H-\overset{..}{\underset{..}{O}}-\overset{:O:}{\overset{\|}{C}}-(CH_2)_{16}-CH_3 \longrightarrow \\ CH_2-\overset{..}{\underset{..}{O}}-H \end{array}$$

$$\begin{array}{l} CH_2-\overset{..}{O}-\overset{:O:}{\overset{\|}{C}}-(CH_2)_{16}-CH_3 \\ \qquad\quad :O: \\ CH-\overset{..}{O}-\overset{\|}{C}-(CH_2)_{16}-CH_3 \quad + \quad 3H_2O \\ CH_2-\overset{..}{O}-\overset{}{C}-(CH_2)_{16}-CH_3 \\ \qquad\qquad :O: \end{array}$$

triglyceryl stearate

Most esters, including this one, may be broken down by treatment with a base such as sodium hydroxide in a process called *saponification*, a term that derives from the role of the reaction in soapmaking. This is the reverse of the ester-forming reaction above.

Saponification reaction:

$$\text{Triglyceryl stearate} + \text{hydroxide ion} \longrightarrow \text{stearate anion} + \text{glycerin}$$

Since the saponification medium is basic, stearic acid is produced as its conjugate base, the stearate ion. A simple proton-transfer reaction interconverts them as shown.

$$CH_3-(CH_2)_{16}-C\overset{\overset{\ddot{O}}{\diagup}}{\underset{\ddot{O}-H}{\diagdown}} + OH^- \longrightarrow CH_3-(CH_2)_{16}-C\overset{\overset{\dot{O}.}{\diagup}}{\underset{\ddot{O}.^-}{\diagdown}} + H_2O$$

When sodium hydroxide is used as the base, sodium stearate is the principal product.

It should be noted that animal fats and vegetable fats such as olive oil also contain other fatty acids besides stearic acid. Palmitic acid has two fewer carbons in the chain. Oleic acid has a carbon–carbon double bond in the chain.

$$CH_3-(CH_2)_{14}-CO_2H$$

palmitic acid, a prime constituent
in palm oil

$$CH_3-(CH_2)_7-CH=CH-(CH_2)_7-CO_2H$$

oleic acid, a prime constituent of olive oil

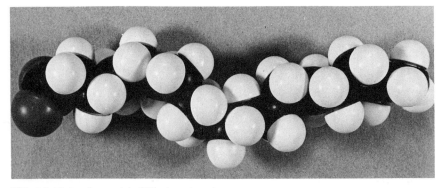

FIG. 2.8 Molecular model of the stearate anion.

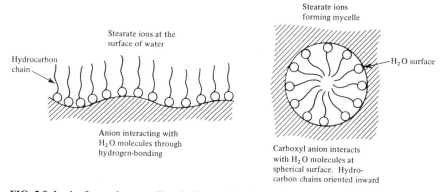

FIG. 2.9 In the figure the wavy lines indicate the hydrocarbon chain and the circles represent the polar carboxylate anions portion.

Soaps perform their cleaning functions by increasing the wetting capabilities of water. They do that because of a "schizophrenic" character of the molecule. The stearate anion has a highly polar area at the carboxylate ion site. This center will coordinate through hydrogen bonding with water molecules. The long hydrocarbon chain will not interact well with water molecules. In fact, this part of the molecule would prefer not to be in the water at all. There are two ways in which both of the conflicting tendencies can be satisfied. One is for the molecule to place itself at the surface of the solution with hydrocarbon portions jutting outward and the carboxylate anion part interacting with water molecules at the surface. The other way is for the hydrocarbon portion to be oriented into the center of the sphere and the anion part interacting at the surface with water molecules. These spheres are called *mycelles*.* The two forms are shown in Fig. 2.9.

In its position at the surface of water, the soap changes the characteristics of the surface by reducing the surface tension; the ability of water to penetrate into particles adhering to a fabric is enhanced. The dirt is thereby "lifted out" by the operation. The slippery feeling of soapy solutions is also due to the surface changes. Because they act at the surface, soaps and detergents are known as *surfactants*.*

There is one very serious problem connected with soaps, and that is the problem of the accompanying cation, which is usually the sodium or potassium ion. When these are replaced by calcium, magnesium, or ferric ions that are frequently dissolved into domestic water supplies from mineral deposits, the soap loses its solubility. Water containing these ions is called *hard water*.

$$2CH_3(CH_2)_{16}-CO_2^- + 2Na^+ + Ca^{2+}$$

$$\longrightarrow [CH_3-(CH_2)_{16}CO_2]_2Ca + 2Na^+$$
calcium stearite precipitate

The stearate salts of these ions are much less soluble in water and precipitation from solution occurs. This both reduces the concentration of the soap and forms a film on the fabric. Thus, the cleaning effectiveness of the soap is doubly reduced in such water.

Two options are available. One is to "soften" the water by removing the offending cations. The other is to use an alternate soap that will tolerate higher concentrations of the hard water cations without losing its cleaning effectiveness. This latter approach led to the development of synthetic detergents shortly after World War II. The first approach is used in commercial and domestic water softener systems that cause the reverse ion exchange, substituting sodium ions for the hard water ions. Such soft water can be used effectively. The hard water flows through an ion exchange resin that contains large quantities of sodium ions. The exchange traps calcium ions in the resin and releases sodium ions. The water, thus depleted of calcium and other hardening ions, is soft.

$$Na^+ \text{ (resin)} + Ca^{2+} \text{ (in hard water)} \longrightarrow Ca^{2+} \text{ (resin)} + Na^+ \text{ (in soft water)}$$

When the sodium ions in the resin are depleted, the resin can be recharged by adding a concentrated solution of sodium chloride. This re-exchanges the hardening ion and sodium ions. The recharging solution is expelled and the resin is ready for further softening operations.

$$Ca^{2+} \text{ (resin)} + Na^+ \text{ (from recharging solution)} \longrightarrow Na^+ \text{ (resin)} + Ca^{2+}$$

in spent softener   recharged

A diagram of a water-softening operation is shown in Fig. 2.10.

SYNTHETIC DETERGENTS

The most important class of compounds used as synthetic detergents are from a group known as *sulfonic acids*. These compounds are similar to the fatty acids of natural soaps in that they are weak acids in aqueous solution. An example is the phenyl compound, benzene sulfonic acid.

benzenesulfonic acid   benzenesulfonate anion

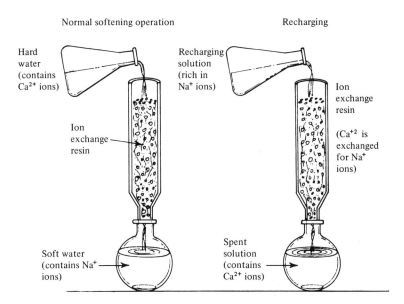

FIG. 2.10 The water softener.

The acid is a somewhat stronger acid than the corresponding carboxylic acid, benzoic acid:

$$\langle\bigcirc\rangle-CO_2H + H_2O \rightleftharpoons \langle\bigcirc\rangle-CO_2^- + H_3O^+$$

If a long hydrocarbon chain is now attached to the benzene ring of the sulfonic acid, the surfactant characteristic that we found for sodium stearate should be present again. An example is sodium p-lauryl-benzenesulfonate with a 12-carbon chain.

$$CH_3-(CH_2)_{10}-CH_2-\langle\bigcirc\rangle-SO_3^- + Na^+$$

The greater effectiveness of this detergent in hard water exists because of somewhat higher solubility of the anion in the presence of hard water cations. But problems of another sort emerged with synthetic detergents shortly after their discovery.

Stearic acid and the other carboxylic acids are consumed by organisms in nature and are used to synthesize fat tissues. This is why they are sometimes called *fatty acids*. They also can be metabolized and are rich energy sources. Thus, the soap effluent can be incorporated readily into natural organisms, especially by microorganisms in water, and degraded eventually to CO_2 and water. The removal of the soap from the environment is thus accomplished without any problems.

On the other hand, the sulfonic acids do not occur in nature, and natural pathways for their decomposition do not exist. Chemical oxidation by molecular oxygen, which is another degradation route, is difficult, hence, the detergent molecules are stable and tend to persist. By the time synthetic detergent use had become widespread in the late 1950's, this problem became apparent because of foaming of water in municipal sewage systems, in streams near municipal sewage treatment outfalls, and even in tap water. This most visible sign of water pollution became a matter of great public concern. Efforts by the detergent industry solved this problem. The synthetic detergents first used were of the general structure where R is the hydrocarbon chain, as with sodium p-laurylbenzenesulfonate.

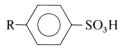

The most widely used synthetics had long chain alkyls of branched rather than linear structure present, however. For example,

$$
\begin{array}{c}
\underset{|}{CH_3} \qquad \underset{|}{CH_3} \\
CH{-}CH_2{-}CH{-}CH_2{-}\!\!\!\bigcirc\!\!\!{-}SO_3^-\,Na^+ \\
\underset{|}{CH_2} \\
\underset{|}{CH{-}CH_3} \\
\underset{|}{CH_2} \\
CH_2{-}CH_3
\end{array}
$$

Research showed that the nature of the alkyl chain was an important factor in biodegradation. If R is a straight chain of carbon atoms as with sodium p-laurylbenzenesulfonate, the compound is biodegradable. If R is a branched chain, the compound is not so readily biodegradable. This structural change was made and reduced the foaming problem.

Another type of biodegradable synthetic detergent is a fatty acid derivative formed by reduction of the fatty acid by hydrogenation. Stearyl alcohol is produced.

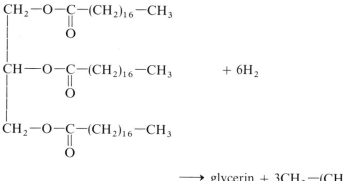

$$\longrightarrow \text{glycerin} + 3CH_3-(CH_2)_{16}-CH_2OH$$

stearyl alcohol

The alcohol forms an ester with sulfuric acid.

$$CH_3-(CH_2)_{16}CH_2OH + H_2SO_4$$

$$\longrightarrow CH_3-(CH_2)_{16}CH_2-O-SO_3H + H_2O$$

The sodium salt of this ester is formed by the removal of the other proton in a proton-transfer reaction with bases such as sodium hydroxide.

$$CH_3(CH_2)_{16}CH_2-O-SO_3H + NaOH$$

$$\longrightarrow CH_3(CH_2)_{16}-CH_2-O-SO_3^- Na^+ + H_2O$$

Again we have detergent properties—a polar group and a long hydrocarbon chain on the same molecule. These are known *linear alkyl sulfates* and are also biodegradable.

The success of biodegradable synthetic detergents is a testimony in support of the school of thought that contends that technologically caused pollution problems may be solved by further technological developments.

PHOSPHATES IN DETERGENTS

The principal phosphate constituent in detergents is sodium polyphosphate, $Na_5P_3O_{10}$, which is more properly named pentasodium triphosphate. Triphosphate is a three-unit phosphate anhydride. The phosphate linkage ties two phosphates from H_3PO_4 together by splitting out water.

Triphosphoric acid,

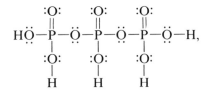

is the parent acid. Replacement of the five protons with sodium ions produces the sodium polyphosphate.

Phosphates in detergents are called *builders*. They have two essential roles. One is to act as a sequestering agent or chelating agent* for metal ions, especially the hard water cations, iron(III) and calcium(II). A chelating agent is one that can form a coordination bond with a cation at more than one site, as shown. Coordination between polyphosphates and calcium ion can occur with two or more oxygen atoms on the phosphate ion. The cation is thus unavailable for interacting with the detergent and so the cleaning action of the detergent is enhanced.

The other purpose of the phosphate is to help make the solution alkaline. The reason for this is to be found upon looking at the acid-base behavior of the phosphate system. Again in water the polyphosphate breaks down in part to form monophosphates or *ortho-phosphate*. This ion will involve itself in equilibria* involving proton-transfer reactions.*

These three acid-base reactions are

$$H_3PO_4 + H_2O \rightleftharpoons H_2PO_4^- + H_3O^+ \qquad K_1 = 7.5 \times 10^{-3}$$

$$H_2PO_4^- + H_2O \rightleftharpoons HPO_4^{2-} + H_3O^+ \qquad K_2 = 6.2 \times 10^{-8}$$

$$HPO_4^- + H_2O \rightleftharpoons PO_4^{3-} + H_3O \qquad K_3 = 2.2 \times 10^{-13}$$

If phosphate is introduced into a solution, it will function as a base, acquiring a proton from water.

$$PO_4^{3-} + H_2O \rightleftharpoons HPO_4^{2-} + OH^-$$

The value of K for the hydrolysis may be determined from our knowledge of K_3 and from the dissociation constant of water.

$$K_{\text{hydrolysis}} = \frac{K_w}{K_3} = \frac{1 \times 10^{-14}}{2.2 \times 10^{-13}} = 4.5 \times 10^{-2}$$

Since this is a fairly large value, it means that considerable hydroxide ion concentration is found in typical detergent solutions because the equilibrium

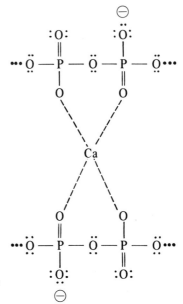

FIG. 2.11 Chelation of Ca^{2+} by polyphosphate.

lies somewhat to the right in the phosphate hydrolysis. They have pH's of 9–10. When the pH is that basic, the detergent will be in the form of its conjugate base rather than the uncharged acid, which is a less effective detergent.

The involvement of phosphates in eutrophication and the fact that detergents contribute significant quantities of the phosphate present in municipal sewage has led some communities to ban phosphates in detergents. When this happens, the options are either to get along without the advantages in effective cleaning that phosphates give a detergent or to find substitutes for phosphates. A third possibility is to use water softeners on a far more extensive level than heretofore. The option that has been studied the most is that of replacing phosphates with another compound or formulation.

Alternatives to phosphates have been developed because of the nutrient problem. One such compound is nitrolotriacetic acid, called NTA.

Note that the compound is a carboxylic acid composed of three acetic acids bonding through the methyl carbons to a nitrogen to form an amine. It is capable of forming chelates with metal cations, such as Fe(III) and Ca(II) ions. The chelating cation of NTA is similar to that of a compound long used for this purpose in analytical chemistry, called ethylenediamine tetraacetic acid (EDTA), which has the structure

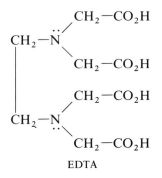

EDTA

Bonding between the cation and the chelator takes place through both the oxygens of the carboxyl groups and the amino nitrogen groups. Both of these sites are Lewis bases* because they are electron-rich centers.

There are still several questions to be answered about NTA. Since the compound would be dispersed into water that may later be consumed, its safety for human consumption must be confirmed. Also, it is necessary to ensure that the biodegradation products from NTA are safe.

Some fear has been expressed that NTA will chelate with metal cations too well. By combining with a toxic metal cation such as mercury(II) or cadmium(II) ions, the effective concentration of these ions in water could be increased. This would probably have no effect on mature humans who consumed the water but it is possible that the chelation could cause these cations to be transported across the placental barrier between a mother and a developing fetus in sufficient amounts to cause *teratogenic* effects (birth defects). There is also the possibility that the critical element in eutrophication is nitrogen for some streams. If so, degradation products of nitrogen-containing NTA would enhance plant growth more than phosphates. These possibilities are of sufficient concern to cast doubt on the future role of NTA, and its use in detergents has been disallowed by the federal government.

Another approach that has recently developed commercially is the marketing of phosphate-free detergents. These formulations generally substitute soda ash (sodium carbonate) and sodium silicates for sodium polyphosphate. Both carbonate and silicate ions hydrolyze in water, in the same manner as phosphates do, producing hydroxide ion.

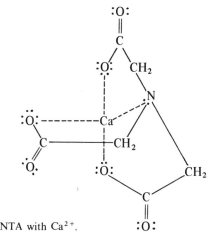

FIG. 2.12 Chelation of NTA with Ca^{2+}.

$$CO_3^{2-} + H_2O \rightleftharpoons HCO_3^{\ominus} + {}^{\ominus}OH$$
$$SiO_4^{4-} + H_2O \rightleftharpoons HSiO_4^{3-} + {}^{\ominus}OH$$

Phosphate-free detergents are usually more alkaline than those containing phosphates. The pH is between 10 and 11 for the detergent solutions. This represents a hydroxide ion concentration between 10^{-3} and 10^{-4} molar, and this causes problems of handling and safety both for the detergent powder and for the solution. Another problem is that the enhanced carbonate concentration serves in the same manner as raw sewage as a source of carbon nutrients through its conversion to CO_2 through two hydrolysis steps.

$$CO_3^{2-} + H_2O \rightleftharpoons HCO_3^{\ominus} + {}^{\ominus}OH$$
$$HCO_3^{-} + H_2O \rightleftharpoons CO_2 + H_2O + {}^{\ominus}OH$$

Thus increased carbonate concentration could enhance plant growth development.

While phosphates in detergents present problems, substitutes may pose yet more serious hazards. Perhaps the other approach—to remove phosphate from waste water instead of substituting something for phosphate—should be considered. Since most of the detergent used is processed through a municipal sewage disposal plant, a logical approach might be treatment of sewage wastes. Before considering that, let us look at the overall problem associated with sewage treatment, since this is another aspect of the nutrient problem.

SEWAGE TREATMENT—THE "FLUSH-IT-AND-FORGET-IT" SYNDROME

The national output of raw sewage from human beings is very great indeed. The total domestic wastes amount to 3 to 4 trillion gallons per year. It contains 9 billion lb of solid wastes. An investment in waste treatment plants of approximately 880 million dollars is made every year. Federal legislation providing 10 billion dollars in additional funds for waste treatment has been enacted. A community of 25,000 will process approximately 1 million gallons of raw sewage per day.

The suspended solid waste in raw sewage, mostly fecal matter, is only about 0.5% of the total by volume. These suspended solids are removed by settling in a pond or lagoon, a procedure termed *primary treatment*. This process removes enormous quantities of organic matter as a sludge, but large amounts remain dissolved. *Secondary treatment* is aimed at removing the organic compounds in solution. The liquid effluent is exposed to bacteria in a settling pond or is filtered through layers of rock or charcoal. The bacteria, as decomposers, oxidize carbon compounds to CO_2. This secondary effluent is now much reduced in organic carbon content, and this is the point at which nearly all sewage treatment operations terminate. Serious problems are still involved, however. The effluent is high in CO_2 content and low in O_2 because of the secondary treatment. Moreover, inorganic products have not been removed at all. Most important of these are nitrates and phosphates, which are present in substantial amounts.

Experimental operations are underway studying the effectiveness of a tertiary treatment by the addition of lime, or calcium oxide. When added to water, CaO forms its hydrated form, calcium hydroxide.

$$CaO + H_2O \longrightarrow Ca(OH)_2$$

Calcium hydroxide is slightly soluble in water, and the presence of calcium ion in the solution will cause the precipitation of phosphate as calcium phosphate.

$$3Ca^{2+} + 2PO_4^{3-} \longrightarrow Ca_3(PO_4)_2$$

In addition, CO_2 is removed by precipitation as calcium carbonate.

$$Ca^{2+} + CO_3^{2-} \longrightarrow CaCO_3$$

Carbonate is present because of the dissociation of CO_2 as a weak acid in water.

The major problem is expense. Tertiary treatment costs approximately twenty cents per thousand gallons at present experimental operations. Part of this cost can be recovered by recycling of the calcium carbonate formed, by heating to drive off CO_2 and recovering lime for reuse.

$$CaCO_3 \xrightarrow{\text{heat}} CaO + CO_2$$

This tertiary treatment substantially removes phosphates. With this operation, the advantages of phosphate as detergent would be possible and the accompanying eutrophication problem would be alleviated.

Some interesting new ideas about sewage treatment and disposal have been studied. One logical approach would be to use the settled sludge or to use the effluent liquid as an agricultural fertilizer on a massive basis. Milwaukee, Wisconsin, is presently experimenting with commercial marketing of its solid wastes as a fertilizer. As one would expect, plots treated with such effluent show significant increases in crop yields, without any danger to groundwater purity or other contamination developing. Another advantage found was that the mineral nutrients deposited on the soil were largely retained rather than running off again. An engineering and economic problem presents itself though. . How is waste sewage to be transported from the urban center to an agricultural area at a reasonable cost?

Another idea is to burn the dry sewage solids in a conventional electric generator power plant. Two problems are alleviated at the same time. The consumption of fossil fuels is reduced, and the sewage is disposed of with minimal ecological disruption. Nothing is done in this process about dissolved carbon or inorganic nutrients. It has also been proposed that we allow for the removal of dissolved organics by letting eutrophication take its course in a closed system, periodically harvesting the algae that are growing. The problem with this system is that no one knows what effective use can be made of algae, nor has the necessary technology emerged.

Despite these possibilities many communities treat their municipal wastes too little or not at all. It is clear that removal of nutrients from this source is technically possible but that conversions to better systems will be a costly process. Unfortunately the need for improvement in municipal sewage treatment has become apparent at a time when local governments are hard pressed to meet present expenses and voters seem uninclined to approve tax increases. The fact remains that pollution control is an expensive proposition.

SUGGESTED READING

1. *Cleaning Our Environment—The Chemical Basis for Action*, American Chemical Society, Special Issue Sales, 1969. A major portion of this reference discusses water pollution related to eutrophication.

2. *Agriculture and the Quality of Our Environment,* Nyle C. Brady, ed., American Association for the Advancement of Science, AAAS Publication 85, Washington, D.C., 1967. This volume contains a number of important papers. Of particular interest to the reader may be "Fertilizer Nutrients as Contaminants in Water Supplies" by George E. Smith; "The Animal Waste Disposal Problem" by E. Paul Taiganides; "Waste Water Renovation by the Land—A Living Filter" by Louis T. Kardos; "Eutrophication and Agriculture in the United States" by Jacob Verduin.

3. "The Battle of Lake Erie: Eutrophication and Political Fragmentation" by Kathleen Sperry, *Science,* **158**, 351–355 (1967). This analysis describes in further detail the problems associated with Lake Erie.

4. *Concepts of Ecology* by Edward J. Kormondy, Prentice-Hall, Inc., Englewood Cliffs, N.J., 1969. This book describes the ecological phenomenon of eutrophication.

5. "Sources of Nitrogen and Phosphorus in Water Supplies," Task Group Report, P. L. McCarty, Chairman, *Journal of American Waterworks Association,* **59**, 344 (1967). An analysis of pollution sources relating to natural water supplies.

6. "Phosphate Replacements: Problems with the Washday Miracles," by Allen L. Hammond in *Science,* **172**, 361 (1971). This article outlines some aspects of the NTA problem and describes other phosphate replacements as well.

3

The Organic Chemicals Industries

Are we now not ready to discuss not only
the gross national product but also the gross national by-product?

John J. Gilligan
Governor of Ohio
(*Address in Cincinnati, Ohio, April, 1971*)

Chemical A major consideration in this chapter centers on the petroleum
Orientation industry. You should review sections in your textbook relating to organic chemistry, in particular, sections on the
chemistry of hydrocarbons, both saturated and unsaturated types. If your
textbook discusses the chemistry of polymers, you should review that area,
and also review sections relating to the chemical industry. Again in this
chapter the subject of acid-base reactions is treated in connection with
chemical processes in pulping of paper, and oxidation and reduction of
inorganic compounds is involved in reactions in waste recovery operations
in the paper industry.

We have seen that much of our pollution problem in lakes, streams, and
coastal estuaries arises from the addition of nutrients by man. This is the
case in varying degree for municipal sewage, detergents, fertilizer runoff, and
other agricultural wastes. In addition to these, the industrial community
contributes a giant share to the effluent burden. A few are responsible for
contributing inorganic nutrients. Fertilizer processing and explosives are
examples of this problem. The more serious problem, however, lies with

those industries that process or produce organic compounds derived from either biological sources or fossil fuels. These industries frequently release organic compounds into streams in large quantity. The subsequent oxidation of these materials to CO_2 by bacterial action, occurring on a large scale, causes the depletion of the oxygen supply in the stream. The stress that this causes on aquatic life can be very serious, in the same way that it does when the oxygen supply is depleted by bacterial action caused by the overgrowth of algae.

A way of determining the extent to which organic compounds are polluting a stream is to measure the amount of oxygen required to oxidize the contaminants completely to CO_2. This quantity is called the *biochemical oxygen demand*,* or the BOD. It is given in units of weight of O_2. The Federal and State water pollution control agencies measure this value by sampling polluted water in order to monitor what is happening. Correlation of oxidizable organic pollutants with their sources is of interest. Output of BOD from all manufacturing outstrips that from domestic sewers threefold.

On a percentage basis, BOD from various industries is shown in Table 3.1. The total BOD output from United States manufacturing operations amounts to about 22 billion lb of oxygen per year. It can be seen that by far the most serious effects come from three industries: the chemical, pulp and paper, and food-processing industries.

TABLE 3.1

STANDARD BIOCHEMICAL OXYGEN DEMAND
CAUSED BY INDUSTRIAL ORGANIC CHEMICAL
EFFLUENTS[1]

Industry	Percentage of Total Industrial Output—Standard BOD (%)
Chemical products	44.2
Paper and allied products	26.8
Food and allied products	19.6
Textiles	4.1
Petroleum and coal	2.3
Primary metals	2.2
Other manufacturing	0.8

[1] *Cleaning Our Environment: The Chemical Basis for Action,* American Chemical Society, Special Issue Sales, Washington, D.C., 1969, p. 97.

PETROLEUM—THE FOUNDATION OF THE ORGANIC CHEMICALS INDUSTRIES

The primary source of industrial organic chemicals is petroleum. There are exceptions to this. For instance, the food and paper industries obviously derive their raw materials from living sources, not fossil fuels. There are a few pharmaceutical products produced by microbial processes. The fermentation industry and soaps and perfumes also are exceptions. A few important compounds are derived from coal. For the rest of the industry the dependence is principally on petroleum.

There is a division of labor within the industry. First we have the petroleum companies, which convert crude petroleum into refined stocks. Their principal objective, of course, is to produce fuels of high performance characteristics, but it is possible and profitable for them to provide raw materials for further refinement as chemicals. At the next level we have the basic chemicals industry whose function is to refine the petroleum product producing bulk chemicals, usually as reasonably pure single compounds. At a third level, the bulk chemicals are then used as raw materials for pharmaceuticals, synthetic fibers, pesticides, and many other consumer products.

There is no fine line of demarcation among these three types of manufacturers. Some companies are engaged at all three levels. Among the 50 largest producers of bulk chemicals, 13 are primarily in the petroleum products field. These producers typically generate 6 to 30% of their revenue from chemical sales. The larger the company, the smaller its percentage of chemical sales tends to be. Giants like Standard Oil of New Jersey and Gulf Oil are relatively low in chemical sales. Smaller companies like Ashland Oil and Cities Service tend to have a higher percentage of their business in petrochemicals.

Among the intermediate bulk chemical producers there is even more overlap. DuPont, Monsanto, Union Carbide, and the other giant chemical producers are involved also in consumer production of chemicals. There is also some interest in this group in expanding their operations at the petroleum refinery level, for the purpose of supplying their own feedstocks for petrochemicals.

The Refining Process

The first step in the refining operation of crude petroleum is to separate the mixture of hydrocarbons on the basis of volatility.* This is done by fractional distillation,* a separation procedure conducted in large towers. Petroleum is distilled by gradually increasing the temperature as the more volatile hydrocarbons are evaporated. The first distillate will be those gases that are dissolved in petroleum. Natural gas is sometimes tapped separately, and it is sometimes included with the low boiling hydrocarbons, the propanes and butanes. The latter may be liquified under pressure and are sometimes known as liquid petroleum gas, or as bottled gas, and it is used as a substitute for natural gas. In higher boiling fractions we obtain petroleum ether, a mixture of the most volatile liquid hydrocarbons. Gasoline distills at 50–200°C and is a complex mixture of hydrocarbons, primarily in the C_6 to C_{12} range. In higher boiling point regions one obtains kerosene, and fuel oil or diesel oil. Materials that serve as waxes and lubricants are obtained from distillation at lower pressures. Waxes and lubricants are hydrocarbons of higher molecular weight and lower volatility. It is impossible to distill them at atmospheric pressure because the temperature necessary to boil them is so high that thermal decomposition would take place instead. To obviate this, they are distilled at reduced pressure, under which condition the boiling temperature is lowered, so that the distillation takes place with little decomposition. What is left at the end of the entire fractional distillation process is a nondistillable asphalt residue constituted of hydrocarbons of very high molecular weight, plus tarry materials that have formed during the earlier heating.

Thus we have a large range of products produced by primary refining operations. Production data for these fractions are found in Table 3.2. One of the difficulties of petroleum refining is the uneven demand for the various products. Until about 1920, gasoline was the white elephant of refining since kerosene was in heavy demand for use in lighting and heating. Now that situation is reversed because of the automobile. The application of further refining is designed to convert the less valuable fractions into the more

TABLE 3.2
ANNUAL U.S. PRODUCTION OF PETROLEUM PRODUCTS[1,2]

	Production (millions of barrels)	Percent of Total
Gasoline	1,704	44.2
Fuel oil	1,034	26.9
Liquified gases	256	6.7
Jet fuel	191	5.0
Still gas for fuel	135	3.5
Asphalt	124	3.2
Petrochemical feedstocks	109	2.8
Kerosene	94	2.5
All others	207	5.2

[1] *Kirk–Othmer Encyclopedia of Chemical Technology*, 2nd ed., **15**, p. 78.
[2] 1965 Production Data.

valuable ones. It is very worthwhile to convert as much of the petroleum ether and kerosene fractions as possible into gasoline.

American refineries can process 11 million barrels of crude petroleum each day. A breakdown of the products derived is shown in Table 3.2.

The next phases of petroleum refining are those secondary processes that both increase the quantity of gasoline recovered from a given amount of petroleum and improve the fuel characteristics of the gasoline. *Cracking* is the term used to describe processes whereby larger hydrocarbons, containing 12 to 20 carbon atoms, are broken down to those of suitable volatility* for gasoline. In this process quantities of ethylene and other unsaturated hydrocarbons or alkenes are produced by loss of hydrogen from saturated alkanes.

The kinds of reactions that occur are typified using a small molecule, isobutane, as an example. Upon loss of H_2, the alkene, isobutene, forms.

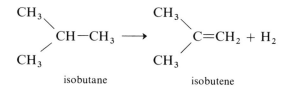

Breakage of a carbon–carbon bond with loss of a hydrogen atom from an adjacent carbon leads to the formation of the C_3-alkene, propene, and methane:

$$\begin{array}{c}CH_3 \\ \diagdown \\ CH-CH_3 \longrightarrow CH_3-CH=CH_2 + CH_4 \\ \diagup \\ CH_3 \qquad\qquad \text{propene}\end{array}$$

With complex molecules, the reactions occurring are similar, but with more possible reaction routes. The net effect is to produce small hydrocarbons from larger ones and to increase the amount of unsaturation present. Some amount of cracking takes place through the heating involved thermally in the distillation. Most, however, occurs because of the presence of solid catalysts, primarily clay derivatives. The cracking catalysts serve as active surfaces to facilitate the cracking reaction when the hydrocarbons are processed at high temperatures.

Cracking leads to large quantities of alkene products, such as isobutene, C_4H_8, and propene, C_3H_6, shown above. Less volatile alkenes are valuable both because they enhance antiknock properties, and because they can be separated and serve as intermediates for further chemical development. Of all the alkenes the simplest, ethylene, C_2H_4, is the most valuable. Production of 128 billion lb of ethylene annually attests to this value. It is the single most heavily used petrochemical. Its principal uses are in the production of polyethylene, the polymeric packaging material, and ethylene oxide, $CH_2 \overline{} CH_2$, used principally in epoxy resins. Ethanol and ethylene

$\ddot{O}$

glycol, a principal component of antifreeze, are also end products from ethylene. In addition, halogen-containing derivatives of ethylene are used as chemical intermediates. Propylene is used to form the polymer poly-propylene as well as other derivatives. Butadiene, $CH_2{=}CH{-}CH{=}CH_2$, is used in preparing synthetic rubber polymers.

A second refining process of great importance is the formation of aromatic compounds from cyclic and straight chain hydrocarbons. An example is the formation of toluene from methylcyclohexane through the loss of one hydrogen atom from each ring carbon.

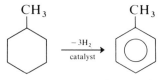

It is possible to obtain methylcyclohexane, in turn, by an isomerization of the linear C-7 hydrocarbon, n-heptane, with loss of one mole of H_2.

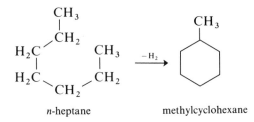

n-heptane   methylcyclohexane

The reaction again is accomplished by high temperature treatment with active catalysts. This route provides toluene and the other aromatics for use in gasoline. The purpose is to increase the octane number of gasoline without the use of lead alkyls (see p. 50). Aromatics are extremely useful as petrochemical synthetic intermediates also. In particular, their use is important in numerous polymers including polystyrene, pharmaceuticals, pesticides, and synthetic fibers. Annual production of aromatics follows in Table 3.3.

Supertankers, Superpollution?

One of the principal problems of the petroleum industry is logistics. Matter has to be moved around, first the crude oil from the field to the

TABLE 3.3
PRODUCTION OF AROMATIC HYDROCARBONS —
USA —1970[1]

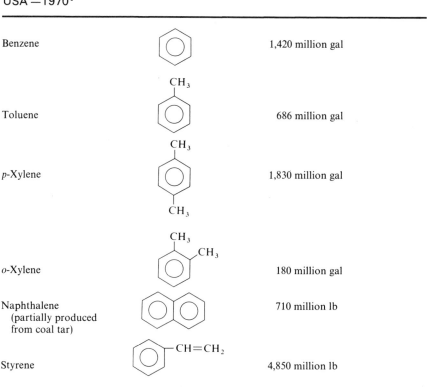

Benzene		1,420 million gal
Toluene		686 million gal
p-Xylene		1,830 million gal
o-Xylene		180 million gal
Naphthalene (partially produced from coal tar)		710 million lb
Styrene		4,850 million lb

[1] H. C. Neely, *Chemical and Engineering News*, **48** (37), 61A (1970).

refinery and then the products to the ultimate consumer. Most of the oil fields now in production are in the Middle East and in Venezuela. Most of, the oil refineries are located in the North Temperate Zones, in Western Europe, and on the East and Gulf Coasts in the United States. Much of the crude oil is shipped for processing in tanker ships. In part due to the closing of the Suez Canal in 1967 and the political instability in the Middle East endangering pipelines to the Mediterranean, the industry has been moving toward the use of very much larger vessels for petroleum transport, especially from the Persian Gulf around Africa. Tankers capable of transporting a half million tons of petroleum are now in operation.

Petroleum discharge from these tankers into the sea occurs both accidentally and through standard operating techniques. The latter have to do with the release of seawater used as ballast when the tanker is traveling without

FIG. 3.1 This super tanker has a capacity of 325,000 tons of petroleum, or more than two million barrels. Its length is nearly half a kilometer. From the point of view of the petroleum industries of the world, tankers of this dimension or larger are necessary economically in order to move petroleum from the producing areas of the Middle East and elsewhere into the energy-consuming developed countries such as Western Europe, the United States, and Japan. (Photo courtesy of Gulf Oil Corporation.)

a load. During loading and ballast discarding, 2 million tons of petroleum are released into the sea every year, 90% through normal operations.

The most spectacular problems have arisen during accidental spills caused by tanker collisions. The *Torrey Canyon* incident in 1967 is one example. A collision in the English Channel released huge quantities of crude oil from a tanker onto the shores of the south coast of England and also onto the French coast. In such an incident marine life in the local areas is usually devastated. Oil penetration into the feathers of aquatic birds makes flight impossible because of added weight. Shellfish are also seriously affected.

There are other ways in which hydrocarbons enter the seas. Leakage and accidents from offshore drilling operations such as that at Santa Barbara, California, in January, 1969, caused similar problems along that coast. Airborne hydrocarbons from auto emissions and other sources enter the ocean eventually through condensation in rainfall. The amount that is contributed in this way in unknown, but it may be sizeable, since about 90 million tons of airborne emissions are produced annually, and while most is oxidized in the air to CO and CO_2, some must reach the sea. Another source of hydrocarbons that is frequently overlooked is from used motor oil and other lubricants. About 500,000 tons of used motor oil is released in this way through sewers into fresh waterways and subsequently into the sea. Discarding of lubricating oils from industry is probably even greater than from private automobiles.

The principal effects of oil in seawater lie in the significant fact of the insolubility of petroleum in water. As a result, and owing to its lesser density, it forms a film on top of the water. To the extent that this changes the surface characteristics of water by changing rates of evaporation and other exchanges with the atmosphere, the oil can cause serious difficulties. Also, the fact that marine life, especially the plants, is concentrated at the surface means that these organisms will be exposed to the oil in considerable amounts.

This can be even more serious when the fact that the oil can serve as a good solvent for other toxic substances is considered. DDT is about 100 million times more soluble in petroleum and other hydrocarbons than in water. There is some evidence that DDT concentrations are very high in oil films and slicks. If so, given the fact that adverse effects of DDT on marine plant life are already known, it means that the organisms are exposed to even higher concentrations of DDT and other chlorinated hydrocarbon pesticides than had been previously thought.

It is important to keep in mind that the problem is not merely the accidental spill, as devastating as that is for a local area. Such damage is reparable after a time as the oil is dissipated and natural processes move to restore life. The more serious problem, in the opinion of marine biologists, is the widespread occurrence of small amounts of petroleum being released by tankers in routine operation, careless discharges from offshore rigs, and effluents from industry and other land-based sources. More widespread and subtle ecological damage may be occurring from such small but persistent oil sources. The obvious solution to the problem is to prevent oil release whenever possible. Prevention of spills and accidents involves the necessity of international cooperation and of greater vigilance in areas of heavy tanker traffic to avoid collision. More care to avoid release in routine operation must be practiced.

Once a large spill has happened, it must be dealt with, with minimum damage to shores and to marine life. Several approaches are in experimental development. The most successful of these appear to be mechanical skimming devices that can remove the surface waters where the petroleum is concentrated. Detergent materials may be used to disperse the oil slick, but most suffer the disadvantage of being toxic themselves at the concentration levels required for dispersal, and the cure is thus as bad as the disease. Another approach, to burn the oil slick, leads to serious air pollution. Another helpful method is to treat an oil slick with microbes that can consume the oil and thence degrade it. Such bacteria exist and probably are involved in normal slow decomposition of oil in the sea.

CHEMICAL MIDDLEMEN—THE CHEMICAL PROCESS INDUSTRY

The business of the chemical process industry is to take petroleum feedstocks, mostly the unsaturated alkenes and aromatics, and, by chemical reactions of a diverse kind, transform them into other intermediates for consumer products. Thus a CPI plant may produce polymer resins from petroleum raw materials; then it sells the resins to a textile manufacturer who will process the resin into fabrics. In some cases the products are directly retailed to the consumer. This is particularly true of plastic films.

A few major industries are prominent consumers of these chemical intermediates. Two of the most important are discussed elsewhere in this book. The pesticide industry produces 1.2 billion lb/year, and detergents are an even larger market, 5.8 billion lb/year. A large consumption comes in the form of paints and lacquers, where organic compounds play important roles as solvents. The consumption of medicinal chemicals is substantial. It is in fact an annual 4 billion dollar business. From a dollar point of view, the most important types of medicinals produced are the antibiotics, tranquilizers, analgesics, hormones, and vitamins and other nutrients. Included are medicinals for livestock as well as for humans. A growing area of importance is that of growth hormones for livestock and poultry. The largest market of all for the chemical industry, however, is the production of polymers.

The Polymer Industry—Macromolecules

Polymers are very large molecules as we have seen in our consideration of the structures of starch and cellulose (p. 21). Yet there exists a pattern in structure, a repetition of features. The polymer is made by repeating simple structure or monomer units again and again. As nature makes cellulose by condensing one glucose molecule after another, so man has learned to make synthetic polymers by linking together appropriate monomers. In listing the momentous events of the twentieth century, the development of polymerization technology must occupy a lofty position. And the source of the monomeric raw materials are petrochemical stocks. Thus the small percentage (Table 3.2) of crude oil converted to petrochemicals is not a true measure of their importance.

One can classify synthetic polymers conveniently according to structural rigidity—thermosetting (rigid), thermoplastic (mobile at high temperatures), and elastomers (synthetic rubbers). One can also classify them according

TABLE 3.4
POLYOLEFINS AND THEIR MONOMERS

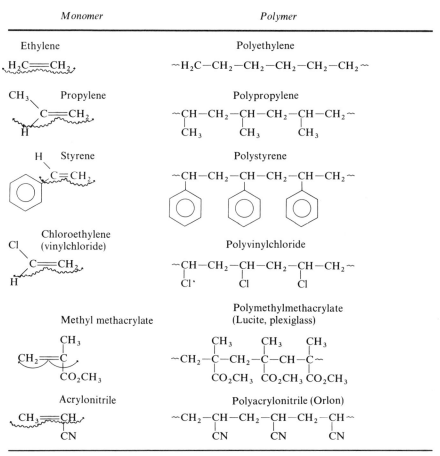

Monomer	*Polymer*

Ethylene / Polyethylene

Propylene / Polypropylene

Styrene / Polystyrene

Chloroethylene (vinylchloride) / Polyvinylchloride

Methyl methacrylate / Polymethylmethacrylate (Lucite, plexiglass)

Acrylonitrile / Polyacrylonitrile (Orlon)

to their monomeric origin. *Polyolefins** are of great importance as structural and packaging materials. The most widely used is polyethylene.

Polyamides and polyesters form the basis for the synthetic fabrics. The basic structural unit is the amide or ester linked together again through repeating chains. Nylon 6 is an example of such a polyamide.

$$\sim(CH_2)_5 - \overset{\displaystyle :O:}{\overset{\|}{C}} - \overset{\cdot\cdot}{N}H - (CH_2)_5 - \overset{\displaystyle :O:}{\overset{\|}{C}} - \overset{\cdot\cdot}{N}H - (CH_2)_5 \sim$$

Its monomer, called caprolactam, is a cyclic amide.

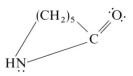

In polymerizing to Nylon 6, the amide bond is formed between molecules instead of within them.

Epoxy, phenolic, and polyurethane resins have widely ranging uses. Epoxy resins are used as adhesives. Phenolics include Bakelite, which is a rigid polymer in which chains are cross-linked to achieve rigidity. Polyurethanes are related to polyesters and are used to form foams. Synthetic rubbers can also be classed as polyolefins except that they still contain many unsaturated linkages. Three examples of their structure are given. Note how the double bonds change position to form links between monomers.

Natural rubber (*cis* polyisoprene):

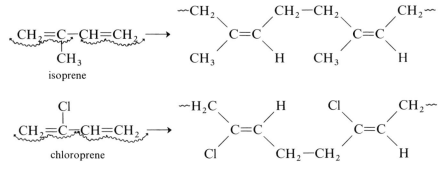

Styrene—butadiene rubber, a copolymer* (two monomers):

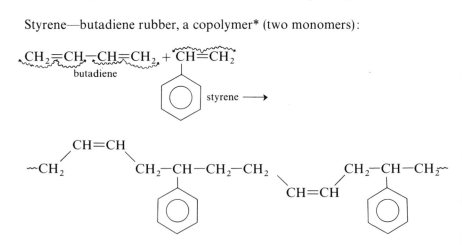

In addition to the basic polymeric material, various additives are included in the formulations. Antioxidants, dyes, and plasticizers, to enhance polymer flexibility, are examples. The output of plastics is truly enormous. About 17 billion lb/year are produced in the United States.

Industrial Synthesis of Organic Compounds

The principal task of the organic chemical industries, then, is to take petrochemical stocks and convert them to organic compounds to be used in formulation of commercial polymers and other compounds. To show how the synthetic chemicals are related to petroleum, let us examine some examples of industrial syntheses.

1. The polyester Dacron:
   The ester linkage in this polymer is formed from two monomers:

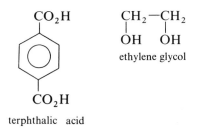

Terphthalic acid is produced from *p*-xylene, a petrochemical (p. 103), by air oxidation in the presence of a catalyst:

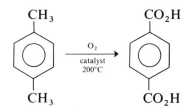

Ethylene glycol is produced from ethylene in two steps. The intermediate is the epoxide, ethylene oxide.

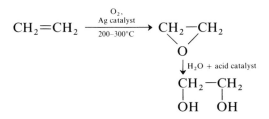

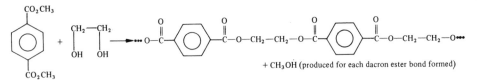

FIG. 3.2 Dacron formation reaction.

In order for the Dacron reaction to occur, terphthalic acid must be converted to its methyl ester, which is more reactive than the parent acid. Reaction with ethylene glycol at 300°C produces a solid polymer that can be drawn into thread (Fig. 3.2).

2. Aspirin is a derivative of salicylic acid:

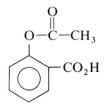

It is the acetate ester of salicylic acid formed at the phenolic OH group:

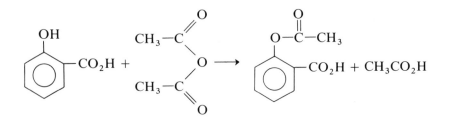

Thus, acetylsalicylic acid is sometimes used as an alternate name for aspirin. Working backward we can trace a relation to benzene and ethylene.

Aspirin forms from salicylic acid and acetic anhydride:

Salicylic acid is formed from phenol by reaction with carbon dioxide in the presence of base:

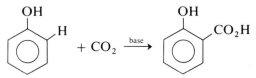

In turn phenol is formed by a reaction between benzene, propylene, O_2, and an acid catalyst in three steps.

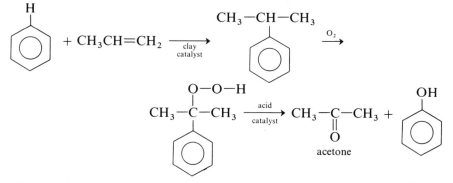

Acetone is very useful in its own right as an organic solvent and synthetic intermediate. It is then a useful by-product in the phenol synthesis. Acetic anhydride is formed from ethylene in three steps:

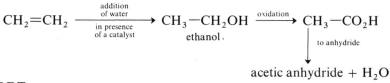

3. DDT:

DDT (p. 128) is formed by reaction of chlorobenzene with chloral in the presence of an acid catalyst.

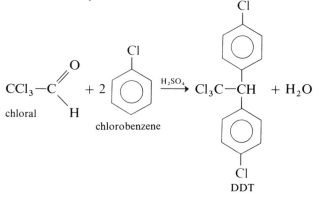

The industrial preparation of chlorobenzene is from benzene in the presence of iron metal as catalyst.

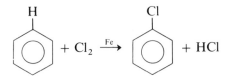

Chloral is produced from ethanol and chlorine.

4. Polystyrene:
    When benzene and ethylene are reacted in the presence of an aluminum chloride catalyst, ethylbenzene forms easily.

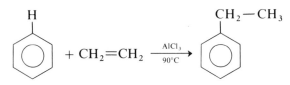

Loss of a molecule of hydrogen, requiring high temperature, forms styrene.

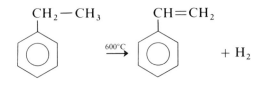

Polymerization occurs readily in the presence of peroxides, which initiate the polymer chain formation at temperatures in the region of 100–200°C. By combining other monomers with styrene it is possible to form copolymers of various structures, such as that shown on p. 108.

CPI and the Environment

The environmental impact of the chemical process industry has two or three very critical aspects. Given the large quantity of materials used in packaging and in fiber, problems associated with disposal have developed. The reason that a synthetic polymer or any other material would constitute a good packaging or fabric material is stability—to heat, light, microbes, or degradation by water or oxygen. Yet when a material no longer serves its original purpose, these very properties mean that it will persist and constitute a littering or disposal problem. Plastic garbage bags persist in a municipal landfill long after organic matter contained within them has decomposed. Some efforts are being made to develop new plastic packaging that can be triggered for decomposition after its intended use has ended. There are other aspects of the disposal problem. For instance, some criticism has been directed at polyvinyl chloride because it gives off HCl gas when heated in an incinerator. This gas present in the effluent is a significant irritant and thus increases air pollution problems associated with municipal refuse incinerators.

Another serious problem is related to a class of compounds known as the *polychlorinated biphenyls*. These materials have been widely used in the organic chemical industry as plasticizers in paints, resins, and plastics. They have high heat capacities and as such are used as heat-transfer fluids and as insulators. They have very low solubility in water, in the parts per billion level.

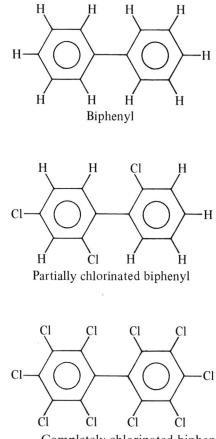

Biphenyl

Partially chlorinated biphenyl

Completely chlorinated biphenyl (PCB) with all ten hydrogens substituted

FIG. 3.3 Structures of biphenyl and chlorinated derivatives.

The ecological effect is that they are rather persistent in nature, tending to accumulate in living matter especially in fatty tissue. In marine diatoms they accumulate to a concentration 1,100 times greater than in the surrounding water. Fish concentrate PCB's from water in the same way. Rats fed PCB in heavy quantities built up much higher concentrations than in the food, especially in fatty tissues and in the liver. There is some slight evidence of inhibition in avian reproduction. The evidence is scanty, in part, because of the fact that we are dealing with a mixture rather than a single compound, since commercial PCB is a mixture of various chlorinated biphenyls. In short, the ecological impact of PCB's is similar to that which we shall shortly find for DDT.

In 1971 a leak occurring in a pipe containing PCB caused contamination of some fish meal being processed. Unfortunately the fish meal was fed to poultry and, as a result of the contamination, PCB levels in the poultry tissues were found to be 22 ppm (parts per million). Since this far exceeded limits on PCB contamination set by the Food and Drug Administration, at 5 ppm, the contamination led to the destruction of nearly 100,000 broilers.

As a result of these findings and concerns, the principal American producer has drastically reduced production of PCB isomers. They are now being produced and used for only those applications in which another material cannot serve the same function. This change in response to scientific evidence and environmental concern probably means that an environmental problem has been solved before it became irreversible and unmanageable and before its impact became too great. A few small companies are still producing PCB's, and no direct action against their use has been taken by environmental agencies.

Another additive group are the phthalate esters. They are used as plasticizers in many polymers to maintain a flexible structure and to prevent crystallization of the polymer.

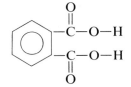

phthalic acid

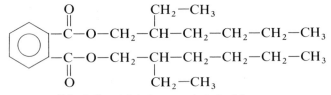

di-2-ethylhexylphthalate, a typical plasticizer

When this ester and similar ones were used in plastics in blood plasma containers, it was found that some of the plasticizer leached into the blood, making the material unsatisfactory for the purpose of blood plasma storage. Similar difficulties with plastic tubing have been reported.

The BOD effluent produced by the chemical process industries is the most serious environmental problem of the industry. It arises from economic factors primarily. In some instances by-products of industrial processes are produced for which there is little demand. It is expensive to treat the material further or to dispose of it, by hauling to a land disposal site in a remote area, for instance, or by combustion to CO_2. The most feasible approach from the point of view of economics is to dispose of it either into the atmosphere or into a stream.

More often the plant itself poses limitations. Even if a company is cautious about effluents, there will be some leakage from the process operations. Valves will leak, chemicals will be spilled, suspended materials will not be trapped properly, and so on. Moreover older plants are not as likely to have state-of-the-art emission control devices as newer ones do, because industry has by and large avoided the problem until recently. In situations where recycling a raw material through a process is profitable, it will be done. At some point, however, another pass will be too expensive, so that the remaining raw material will be otherwise disposed of. Nevertheless the record of the chemical industry in effecting maximum possible recovery is a fairly good record. Like all industry they are reluctant to bear costs they believe too high.

There are some air pollution concerns associated with refining and chemical processes as well. Particulate emission, SO_2, and hydrocarbons all are occasionally problems. Ethylene and other olefins pose a unique air pollution problem in that they serve as inhibitors of plant growth if large enough concentrations are present.

THE PULP AND PAPER INDUSTRY

We consume about 50 million tons of paper and paperboard products each year. This amounts to about a pound and a half per person per day for everything from egg cartons to *Playboy* magazine. Along with a versatile and useful product there are some serious environmental stress factors involved in the pulp and paper industry. First, large consumption of the raw material, primarily wood, is required. The drain on the forest resources can be overcome by proper management. Replacement of trees cut replenishes the forest supply and prevents the secondary effect of soil erosion. Second, some of the chemical processes converting wood into paper pulp lead to serious air pollution in the vicinity of the pulp mill, and perhaps most serious is the magnitude of the biochemical oxygen demand produced by wastes released into streams during the pulp and papermaking process. Lastly there is the disposal of paper after use. The paper industry can and does use waste paper effectively in many applications. About one-fifth of its raw material is paper being recycled, mostly waste from large consumers such as newspaper and book publishers. What is not recycled is biodegradable, almost completely, and that is a distinct advantage. Yet the problem of disposal of paper wastes remains, and it is one of the worst aspects of the solid waste crisis.

After discussing some fundamental aspects of the industry, we shall discuss the air and water pollution problems. Disposal of paper will be discussed in Chapter 6 as it relates to the solid waste problem.

Paper—A Nearly Pure Compound

Paper is mainly cellulose, a single substance, and pulping is the procedure whereby the constituents in wood or other vegetable matter that are not cellulosic are separated out. To the extent that the removal of the other components is incomplete or that other substances are added to the cellulose to convey to the paper some particular characteristic, the paper is not a pure substance.

Any fibrous plant matter is theoretically acceptable for papermaking. Wood comprises about 70 % of the raw material and about 20 % comes from waste paper. The rest is from rags and other plant matter. Both coniferous trees, such as the pine and spruce, and hardwoods are used.

In addition to cellulose, wood contains two other major components. One is the reservoir of other carbohydrates including starches, simple sugars and polymers smaller than cellulose. The other principal component is lignin, which is the adhesive in wood holding the fibers of cellulose into a rigid shape. In dry wood, cellulose constitutes about 60 %; lignin, about 30 %; other sugars, inorganic salts, and resins, 10 % of the weight of the wood. The percentages vary a little from one species to another.

The chemical structure of cellulose was shown on p. 21. It is a polymer of glucose. It will be recalled that glucose in a cyclic form generates a polymeric linkage when two hydroxyl groups condense splitting off a molecule of water. Cellulose has a high affinity for water because it can form hydrogen bonds through the hydroxyl groups, but because of the enormous size of the molecules it is not soluble in water. The polymer chains are also stable in basic media, which is an important factor in the pulping operation.

Lignin is also a polymer, a very complex one, whose structure is incompletely understood although it has been under investigation for more than 50 years. The molecule is known to contain aromatic rings with hydroxyl(-OH) and methoxyl(CH_3O-) groups. These appear to be bound together through carbon chains which contain other functional groups (Fig. 3.4) which form linkages between the monomeric lignin units.

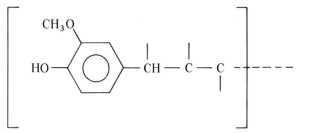

FIG. 3.4 Partial structures of lignin. (Vertical lines show connections to other monomer units.)

The Chemistry of Pulping and Papermaking

The basic process involves four steps:

1. Breakdown of wood into fine particles by mechanical means.
2. Chemical treatment dissolving lignin and other noncellulosic compounds producing paper pulp.
3. The arraying of the resulting pulp slurry of cellulose in water into a moving mass, which is dried, forming a mat of cellulose fibers into paper sheets on the paper machine.
4. The addition, at some point in the process, of other substances, such as dyes, coating materials, and preservatives, depending on the end use of the paper grade.

There are several approaches used in the pulping step, and it is at this point that environmental problems most frequently occur. Of the various approaches, the simplest is to ignore the lignin problem entirely. The wood is ground into fine particles that are suspended in water and matted together to form the sheets. The product quality from the groundwood process is low because the fibers are short, which weakens sheet strength. Also the paper is discolored because of the impurities. For short-lived paper, such as newsprint, where low cost is more important than quality and where neither strength nor surface quality is important, groundwood pulp is the preferred method.

The soda process and the sulfate or Kraft process both involve dissolving the lignin in an alkaline solution. In each case the lignin polymer is broken down and the fragments dissolve by proton transfer from the phenolic hydroxyl group of lignin to the hydroxide ion. At the same time water-soluble compounds in wood are dissolved leaving a cellulose pulp.

FIG. 3.5 A continuous sheet of paper twenty feet wide and over 500 miles long can be produced on this paper machine each day. The paper produced on the machine is used for offset printing, business forms, tablet, and in the manufacture of envelopes. (Photo courtesy of Champion Papers, Courtland Mill, Courtland, Alabama.)

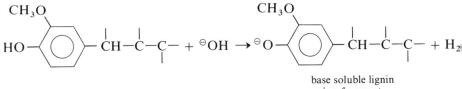

base soluble lignin
anion fragment

The oldest such chemical process is the soda process. Once widely used, it has now been largely supplanted, but it does illustrate the efforts of the industry to recover a certain amount of waste products. Sodium hydroxide is used as the base. The pH of the cooking liquor is 13–14 (between 0.1 and 1.0 M in ^-OH) and thus distinctly alkaline. The cooking is carried out in a closed digester at high temperature and pressure (170°C, 100–110 psi) from 3 to 8 hours. During the cooking, pressure is occasionally relieved by the release of gas, mostly CO_2, to the atmosphere. The pulp is filtered from the liquor, washed, and then bleached with chlorine.

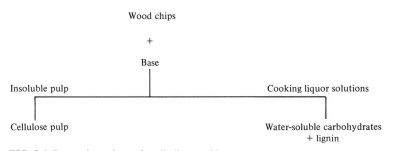

Wood chips

+

Base

Insoluble pulp   Cooking liquor solutions

Cellulose pulp   Water-soluble carbohydrates
                                                  + lignin

FIG. 3.6 Separation scheme for alkaline cooking process

The liquor is recycled to recover sodium hydroxide, because it is a relatively expensive chemical. Water is removed by evaporation leaving liquor solids containing organic residues and inorganic sodium salts. Most of the sodium hydroxide has reacted with CO_2 during cooking, forming sodium carbonate:

$$2Na^{\oplus} + 2^{\ominus}OH + CO_2 \longrightarrow 2Na^{\oplus} + CO_3^{\ominus\ominus} + H_2O$$

After the organics in the liquor have been burned away, sodium hydroxide is recovered by treatment with lime in a process called *recausticization* (regenerating the base).

$$CaO + H_2O \longrightarrow Ca(OH)_2$$
$$\text{lime} \qquad\qquad\quad \text{slaked lime}$$

$$Ca(OH)_2 + Na_2CO_3 \longrightarrow CaCO_3 + 2NaOH$$
$$\qquad\qquad\qquad\qquad\qquad \underset{\text{ppt}}{\text{limestone}} \quad \underset{\text{recycled}}{\text{recovered and}}$$

The limestone precipitate is recycled to lime by heating.

$$CaCO_3 \xrightarrow{\text{heat}} CaO + CO_2$$
$$\qquad\qquad\qquad \text{lime (recycled)}$$

The sulfate or Kraft process includes sodium sulfide in the cooking liquor along with sodium hydroxide. The purpose of the sulfide is to provide a reserve of base for the reaction. Sulfide ion hydrolyzes in water to form hydroxide and hydrogen sulfide ions in a reversible proton transfer:

$$S^{\ominus\ominus} + H_2O \rightleftharpoons HS^{\ominus} + {}^{\ominus}OH$$

Initially, with high concentrations of base, the equilibrium lies to the left because the hydroxide ion is present in large quantity, giving rise to a common ion effect. During cooking, hydroxide ion is consumed and the equilibrium reverses, regenerating hydroxide. Thus the delignification occurs at a more even rate during the entire cooking period, instead of very fast at first and very slow at the end as it does in the soda process. Also the $SH^\ominus$ ion reacts with lignin and enhances its solubility. For this reason delignification is enhanced. Relief of gas from the digester during cooking releases various sulfur-containing gases.

Recovery of chemicals from the spent liquor is similar in the Kraft and soda processes. Sulfide ion is largely oxidized to sulfate in the digesting operation, and this is reversed to recover sulfide. The reduction back to sulfide ion is accomplished using charcoal (carbon) as the reducing agent:

$$SO_4^{\ominus\ominus} + 2C \longrightarrow S^{\ominus\ominus} + 2CO_2$$

Recausticization with lime regenerates NaOH as in the soda process.

In the *sulfite process*, the dissolving of lignin is facilitated by reaction of lignin with hydrogen sulfite ion, HSO_3^-, which forms a sulfonic acid upon reaction with phenolic OH groups. This makes the lignin more soluble in water.

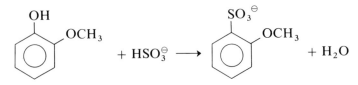

The raw material for hydrogen sulfite ion is sulfur, which is oxidized,

$$S + O_2 \longrightarrow SO_2$$

added to water

$$SO_2 + H_2O \longrightarrow H_2SO_3$$

and made basic to form hydrogen sulfite

$$H_2SO_3 + {}^\ominus OH \longrightarrow HSO_3^\ominus + H_2O$$

Regeneration of cooking chemicals is carried out in the sulfite process in some mills, but not to the extent practiced in the Kraft industry.

Of the two most commonly used processes, Kraft and sulfite, it is ironic that each in a different way affects the environment adversely. With the former the problem is air pollution, arising mainly from release of sulfur-containing gases during digestion and during burning of liquor solids. Among these gases are hydrogen sulfide, H_2S, the "rotten eggs" product of many a famous freshman chemistry lab experiment. Methyl derivatives of H_2S, methyl mercaptan (CH_3SH), and dimethyl sulfide [$(CH_3)_2S$] are also present and so are dimethyl disulfide ($CH_3-S-S-CH_3$) and SO_2. Methyl mercaptan has an even worse odor than H_2S. As well as being unpleasant in odor, these compounds are toxic, but fortunately levels of detection of H_2S are much lower than levels where one experiences noticeable physiological impairment. As a result, people are seldom killed by H_2S fumes because they are aware of them at concentrations sufficiently low to enable corrective action, either elimination of the source or flight. The threshold of odor detection for H_2S is approximately $10 \, \mu g/m^3$ or 7.2 ppb (parts per billion). Severe irritation is manifested at about 20 ppm, more than a thousandfold increase from the detection limit. Death occurs upon exposure to about 400 ppm. Dimethyl disulfide has a similar adverse affect to our olfactory sense. Methyl mercaptan is even lower, with detection at 1 to 2 ppb. The concentration of H_2S from a digester vent may be as high as 19,000 ppm, and the others may be present in the same concentration, or more. Downwind from the mill, the concentration will often exceed thresholds of detection, even after dilution from these high concentrations in the air. The problem is especially intense when the air is still and warm.

The air pollution problem associated with the Kraft pulping process is serious from the point of view of the unpleasant air, although concentrations extant are well below levels thought to cause adverse health effects. It is possible to remove much of the sulfur gases, and equipment to do this is being used with new digesters. Unfortunately, it is not generally being installed in the older Kraft plants. The problem is still very difficult because so little gas can be detected in the air. In addition to the unpleasant odor problem there are some adverse effects on paints especially those containing Pb^{2+}, which forms lead(II) sulfide when reacted with H_2S (p. 176). Lead sulfide is black and so the paint becomes discolored.

In those sulfite mills where there is no recovery, the problem is that of the spent liquor. In the past this was simply released to a nearby stream because no chemicals were recovered, as was the case with the other pulp treatment methods. The sulfite process typically releases 700 lb BOD/ton of pulp produced; the Kraft process produces only 40 lb BOD/ton of pulp. Instead of releasing the effluent to a stream, it can be shunted into an oxidizing lagoon in which the organic material is oxidized in the presence of microorganisms, or to a sewage treatment facility. Another alternative is concentration of the liquor followed by burning. Difficulties associated with air pollution can arise then. Air pollution associated with power generation usually accompanies pulp and papermaking since there are usually power plants tied into the operation. With the sulfite process, there is still resistance in many instances to water pollution limits. Part of the problem is that little commercial use for sulfite liquor chemicals has been found. Lignin is potentially a great source of organic compounds, especially aromatic compounds. Sulfite mills could produce 60 million tons of chemicals per year. The problem is that this source is unable to compete economically with petrochemicals.

THE FOOD PROCESS INDUSTRIES

When Great Grandmother wanted to bake a cake, she probably began by having some wheat ground by the local miller. Grandmother began with prepared flour, and Mother began with a cake mix. Today's housewife may well use a frozen prebaked cake. This example is typical of great change in the eating and living habits of Americans. As farmers' sons and daughters moved into the urban center, an industry grew up concerned with processing foodstuffs and transporting them from the farm to the city. With the increased involvement of women in the work force and professions instead of the kitchen, the degree of prepreparation of foods has increased. For instance, in the last 10 years the per capita consumption of fresh vegetables declined while frozen and canned vegetable usage increased sharply. The tendency for less home preparation of food has led as well to rapid growth in the restaurant industry, the ultimate in processed foods.

The food process industry in all its components does a 115 billion dollar business each year, as Americans spend one-sixth of their income for food. Pollution problems arising from wastes produced during the operations of the industry amount to 4,300 million lb of BOD annually (p. 98), behind only chemicals and papermaking. The food wastes are those portions of the raw foods for which there is little demand or for which further recovery of food material is not economical. Instead of being used to feed people, pets, or livestock, they feed the aquatic organisms in a stream thus depleting the oxygen supply.

Much of the food wastes are in fluids that are given off from various operations. In many cases they contain valuable nutritional substances. For instance, in the production of cheese from milk, about 70% of the nutrient protein in milk is suspended as whey, which is largely discarded. Total production of fluid whey in the world exceeds 50 billion lb/year. A similar loss of protein in blood occurs. Recovery of the protein from these fluids is too expensive to merit doing on economic grounds. Evaporation of the water and recovery of dry solids is expensive because of heat requirements. Alternatively, research is underway to devise a system of removal of the finely divided solids by filtration and by osmotic processes through membranes to concentrate whey solids.

FIG. 3.7 This photograph demonstrates the magnitude of food preparation operations which one encounters routinely in modern food processing plants. Each of the sixteen cylinders is used to cook more than one ton of corn at a time in the preparation of corn flakes. (Photo courtesy of Kellogg Co.)

Still, one could prevent pollution by the nutrients without recovery for their food value, by adsorbing them on catalyst surfaces, for example. If nothing else, solids may be destroyed by incineration. Another approach is to try to recover chemicals from the food wastes. The conversion of waste animal and vegetable fats into soaps and other products is a long-standing example of this. So is the production of the compound furfural, an important commercial aldehyde, from corncobs and other grain wastes:

furfural

An optional use for grain hulls and other vegetable wastes is to heat them in the absence of air. Under these conditions the organic compounds decompose, with reduction to carbon rather than oxidation. The principal product is activated charcoal, which is valuable as a surface-active catalyst in industry. Yet another approach is to treat food wastes by secondary sewage treatment, that is, by bacterial oxidation. This is often a feasible procedure especially for large plants. With small plants the discharge of fluid wastes into a municipal sewage treatment plant with secondary facilities is sometimes possible.

The problems associated with food processing and air pollution also involve the generation of odor in the air from packing plants or other industry. The most pronounced offenders are meat and poultry rendering operations and fish processors. Manure in holding pens in stockyards is also a problem. In rendering, those parts of the animal unsuitable for human or pet consumption are processed. Solid matter such as bone is ground into particles and then with other wastes is heated at 150°C to remove all the water and greases and kill bacteria. What is left is a fine meal that is then used in livestock feed. Among chemicals causing difficulty are the volatile amines (such as trimethyl-amine) that are formed in decaying flesh.

Since dead animals are frequently rendered along with meat or poultry process wastes, the odor problem is compounded. The processing of fish to fish products and fish meal causes the same problems, essentially.

There are two other aspects of importance in the food industry that we shall discuss later. They are problems associated with food additives and with the possibility of toxic effects due to the presence of harmful bacteria.

SUMMARY

The central problem of this chapter has been problems of release of material capable of being oxidized to CO_2 in a stream, causing oxygen depletion and consequent ecological breakdown. To some extent the same approaches can be taken to solve this problem in each industry we have discussed, and also in the pharmaceutical, textile, metallurgical, or any other industrial operations producing organic effluents.

Let us look more broadly at the solutions. The mildest is to try to minimize the difficulty by changing operational procedures. For instance, in batch processes the release of waste into a stream could be done gradually instead of all at once. They may reduce oxygen demand to more tolerable levels.

Strict enforcement of in-plant procedures to minimize spillage and similar actions can reduce effluents but not really very much. The next most rigorous method is to attempt to maximize recycling, even when it is economically unfeasible. An industry will tend to resist this approach unless the added cost is somehow subsidized.

If the wastes cannot be recycled, then it is better to make them less harmful by reacting the wastes with air or, even more effectively, with oxygen. Extensive aeration of an effluent suffices in some instances to speed up the process of bacterial oxidation and at the same time restores oxygen.

Yet a more rigorous approach is to require that wastes, particularly solids, be disposed of in some other way than into a stream. In this category is land disposal, which can provide nutrients for crops, as an additional benefit. The use of in-plant waste treatment facilities is another way. For small industries located near one another, a common waste nutrient treatment facility sometimes operated by a separate company may be the answer. Incineration is also a solution, but of course it is subject to air pollution concerns.

The devising of new sources for waste utilization in products is an area in which industry has a good record when the economics favor utilization. One of the greatest aids for pollution control would be the development of uses for lignin in paper waste liquors. Because there is no use now, lignin wastes are largely discarded.

How much should we blame industry for pollution? In terms of output of biodegradable organics and other pollutants, their contribution is very great. On the other hand, numerous indications are that they are beginning to respond to the demonstrated need for abatement. For instance, consider these relative investments for pollution control equipment in Table 3.5.

TABLE 3.5
INVESTMENT IN POLLUTION CONTROL BY MAJOR INDUSTRIES[1]

Industry	1968	1969	1970	Percent of Net
		(millions of dollars)		Income, 1970
All manufacturing	289.5	394.0	487.2	1.7
Chemical and allied products	50.3	59.3	95.7	2.8
Petroleum and coal products	85.9	89.8	96.7	1.6
Rubber products	7.6	9.7	9.5	0.2
Paper and allied products	26.5	44.2	100.7	14.1

[1] "Facts and Figures in the Chemicals Process Industry," *Chemistry and Engineering News,* **49**(37), 38A (1971).

We see large increases in expenditures particularly by chemical plants and paper mills in recent years.

When we compare data for net income with capital expenditures for pollution (right-hand column of Table 3.5), the performances are less impressive except for the paper industry, which appears to be making a worthy effort to overcome its problem. One must be careful in such generalizations. Some companies in each industry are doing better with pollution abatement practices than others.

SUGGESTED READING

1. "Cleaning Our Environment—The Chemical Basis for Action," American Chemical Society, Washington, D.C., 1969, p. 95–106. This reference discusses the BOD output by various industries.

2. N. Allinger, *et al.*, *Organic Chemistry*, Worth Publishers, Inc., New York, N.Y., 1971, Chapters 25 and 35.

3. A. L. Waddams, *Chemicals from Petroleum*, Chemical Publishing Company, New York, N.Y., 1969. This contains a survey of the petrochemicals industry.

4. For more information on polymer chemistry, see C. T. Greenwood and W. Banks, *Synthetic Higher Polymers*, Oliver & Boyd, Ltd., Edinburgh, 1968. This is a discussion of the subject for newcomers to the field.

5. R. B. Seymour, *Introduction to Polymer Chemistry*, McGraw-Hill Book Co., New York, N.Y., 1971. This is a more advanced treatment of polymers.

6. E. Sutermeister, *The Story of Papermaking*, S. D. Warren Company, Boston, Mass., 1954.

7. J. B. Collerin, *Modern Pulp and Papermaking*, Reinhold Publishing Corp., New York, N.Y., 1957. References 6 and 7 have to do with basic considerations in the pulp and paper industries.

8. *Handbook of Food and Agriculture*, Fred C. Blanck, ed., Reinhold Publishing Corp., New York, N.Y., 1953. This article contains basic information in this field.

9. "Facts and Figures for the Chemical Process Industry." Each year the American Chemical Society publishes extensive data on the chemical process industries and others discussed in this chapter. This information appears each year in an issue of *Chemical and Engineering News*.

10. Marshall Sittig, *Organic Chemical Process Encyclopedia*, 2nd ed., Noyes Development Corporation, Park Ridge, N.J., 1969. This volume describes the chemistry and processes involved in many industrial syntheses.

4

Competition to Man from Other Species — The Pesticide Problem

Perhaps with the new ecological concern,
people would be willing to eat some blemished apples or oranges
or green vegetables with a few holes in the leaves, if they knew
that doing so would mean lower doses of
insecticides released into the environment.

F. R. Lawson, Entomologist
United States Department of Agriculture
(*from* Environment *Magazine, May, 1971*)

Chemical Orientation Most pesticides that are used widely are organic compounds and involve several aspects of organic chemistry. In particular, the reader should be familiar with some of the important classes of organic compounds, including organic halogen compounds, carboxylic acids and esters, benzene derivatives, epoxides, and phenols. A discussion of naming organic compounds is also included in this chapter.

PERSPECTIVE ON PESTICIDES

Like all other living things, man has had to contend with natural enemies. Because of his intelligence, he has seen that these include not only those who prey physically upon him, such as the hookworm or malaria-causing microbes, but also predators upon his domesticated animals and upon his food and fiber supply. Perhaps this last is the most important category of all because it includes so many species. The problem is compounded by modern agriculture, in which large areas are devoted to a single crop. In the wild, individual plants are dispersed and competing with many other plant species.

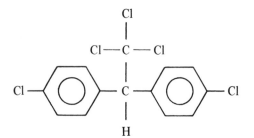

FIG. 4.1 The chemical structure of DDT. In the model the large gray atoms are the chlorines. One chlorine of the CCl_3 group is obscured by the rest of the molecule. The white atoms are hydrogen; the black atoms are carbon.

They are more likely to be separate from their own kind, which means a predator of a given plant species is hard pressed to find his prey. But an initial small infestation by an insect or other predator in a cultivated field will lead to a large increase in their number through reproduction because their food supply is so plentiful there.

In the middle of the twentieth century, the use of chemical agents to control these competitors and enemies of man became widespread. The term *pesticide* (pest killer) describes their general function, and terms like *insecticide, fungicide, herbicide, rodenticide*, etc., refer to pesticides directed to specialized targets or categories of pests.

As the use of pesticides became massive and spread through the world, problems associated with their use emerged in some cases. Resistance to the pesticide developed in some targets. Some of the pesticides eliminated beneficial species or caused bizarre side effects. Handling and application posed hazards to human beings. Now we find serious questions posed about whether some of these substances should be used at all. To understand the problem we first explore in depth that substance that most typifies the problem—both in terms of the benefits and the hazards. The common name of the compound—<u>d</u>ichloro<u>d</u>iphenyl<u>t</u>richloroethane has been abbreviated to the three underlined letters in its name—DDT.

It is not a naturally occurring substance, and so its synthesis from other materials by chemical reaction was required. The first preparation of DDT in the laboratory was carried out in the nineteenth century. The insecticidal properties were neither sought nor found at that time. A second preparation of the substance in 1939 by Paul Mueller, working for a chemical company in Switzerland, led to the discovery of the toxic effects to insects. World War II accelerated its development to control pests as a means of sanitation control for armies in the field.

The industrial preparation of DDT from inexpensive starting materials is possible, and thus its availability on a massive scale was assured. The compounds required are chloral, chlorobenzene, and a sulfuric acid catalyst (p. 111).

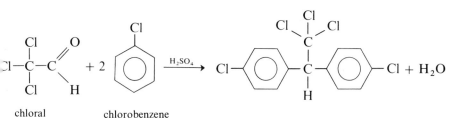

chloral   chlorobenzene

An important side reaction forms an isomer of DDT in which one of the phenyl rings attaches to the *ortho* position, adjacent to the chlorine. Other isomers are also produced in lesser amounts, and commercial DDT is a mixture of these various compounds. It is cheaper to use this mixture than to try to obtain a more potent insecticide by separating out the pure material.

FIG. 4.2 op′-DDT, an isomer of DDT. Note that one of the phenyl rings is bonded to the ethane chain at the adjacent *ortho** position to the chlorine rather than across the ring at the *para** position. There are three positions for a second substituent in a benzene ring with respect to the first. They are the ortho, *meta,** and para positions, and these are shown on the ring at the right.

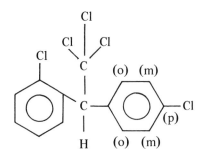

DDT affects the nervous system of insects. The insecticide is ingested either by absorption upon contact or by the insect eating it. After a short period, coordination problems develop and then intensify, leading to convulsions and death. It is thought that DDT is concentrated in lipid* tissues, including the nervous system. A lethal dose for the anopheles mosquito larva is about 5 ppb in their tissues.

The early experience with DDT made it seem to be a panacea. As an insecticide it was ideal. Its greatest toxicity was toward the most serious insect enemies, and beneficial insects such as honeybees seemed less affected. There was little or no short-term problem for human beings who handled the substance. Its use in control of crop pests and in control of the anopheles mosquito skyrocketed, first in the developed and then in the emerging nations.

Production of DDT in the United States shows its history clearly. From the beginning in 1944, production increased from 4,850 to 44,000 tons in 1951. Thereafter, production increased less rapidly, but steadily, to 70,000 tons in 1963. Production of DDT has decreased some in the United States since that time, to about 20,000 tons in the early 1970's, when its domestic use was banned by the Environmental Protection Agency. The decline in the decade of the 1960's is the result of questions raised about the safety and effectiveness of DDT and because it has been replaced by other insecticides. However, the total worldwide production of DDT from its inception to 1974 is estimated to be 3 million tons.

Two properties of DDT are of special importance in the unfortunate environmental side effects that have arisen from its use. The most important

TABLE 4.1
SOLUBILITY OF DDT IN VARIOUS
SOLVENTS[1]

Solvent	Solubility* (g/l of solvent)
Highly Polar	
Water	10^{-6}
Triethanol amine	10
Propylene glycol	10
Moderately Polar	
Ethanol, 95%	20
Isopropyl alcohol	30
Acetone	580
Nonpolar	
Gasoline	100
Carbon tetrachloride	450
Chlorobenzene	740
Benzene	780

[1] Donald E. H. Frear, *Chemistry of the Pesticides*, D. Van Nostrand Co., Inc., Princeton, N.J., 1955.

molecular feature is its lack of solubility in water. This is a typical characteristic of chlorinated hydrocarbons generally. It arises because of the failure
of the molecule to interact favorably with water molecules either through
hydrogen bonding or other dipolar interactions. As a result the solubility
of DDT in water is only 1 ppb or 10^{-6} g DDT per liter of water at 30°C.
In solvents of low polarity, the solubility increases, as shown in Table 4.1.

Because of this disparity in solubility, DDT will tend to prefer some living
tissues in preference to others because of differences in polarity in the tissues.
In particular, DDT will concentrate in tissues containing large amounts of
fats, apparently because these tissues have high concentrations of compounds
of relatively low polarity. Highest concentrations are found in adipose, or
fatty, tissues; in liver and brain tissues; and also in the gonads. The fat
content of mammals' milk causes DDT concentrations to be somewhat
higher there. A preference for similar tissues in plants is also manifested.

The second chemical factor of importance is the resistance of DDT to
decomposition once it has been released into the environment. The normal
breakdown pathways of photochemical decomposition (interaction with
light), oxidation by O_2, or microbial breakdown do not affect DDT. The
persistence of DDT has been measured and is considerable. After a single
application some DDT is still detectable in soil 3 years later. At one time this
persistence was considered a virtue because it meant that a crop could be
protected through its entire growing season instead of just for a few days after
spraying.

While some DDT is retained, much of it is removed from the original site
of application. It is carried off in water, evaporated into the air, and incorporated into living plant and animal tissues. The tendency is for these
factors to cause the pesticide to be transported through the environment.
Since it is stable, it persists and has become widespread. Detectable amounts
of DDT have been found in Antarctica and in the Central Pacific, thousands
of miles from the nearest site of application. DDT is everywhere.

Some mechanisms for the breakdown of DDT do occur. The most
prominent is the loss of HCl—dehydrochlorination. The resulting product
contains a carbon–carbon double bond, which makes it a derivative of
ethylene. The resulting product is named dichlorophenyldichloroethylene or
DDE.

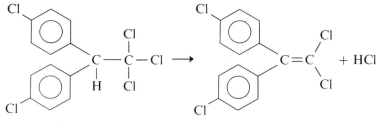

loss of HCl   DDE

DDE itself has insecticidal properties, although it is not as potent as DDT. This is not a particularly rapid reaction, and DDE suffers the same difficulty of persistence in the environment as DDT.

But the problem does not end with the persistence factor. DDT and other chlorinated hydrocarbons have a special affinity for living things. For instance, studies of DDT concentration in the sediments of a portion of Lake Michigan revealed 0.0085 ppm of DDT. In invertebrates such as crayfish taken from the same area, a concentration of 0.41 ppm was detected. The fish predators of these invertebrates had concentrations of about 5 ppm. Herring gulls, which feed on these fish, had concentrations of 3,200 ppm in their fatty tissues. Humans carry 5 to 20 ppm of DDT and analogs in body fat, while our food only contains 0.050 to 0.50 ppm.

As one goes up a series of predators in a food chain, the concentration of DDT is strongly increased. So, while levels of DDT at the lower end of the food chain may be well below the danger zone, the higher carnivores may face serious difficulty.

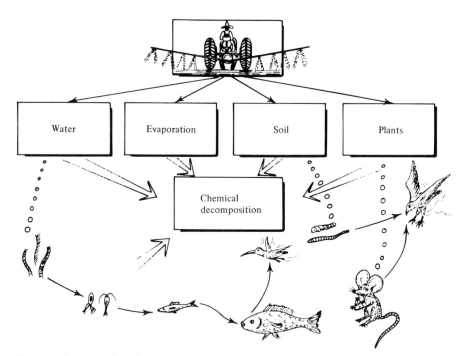

FIG. 4.3 The interaction of a stable pesticide such as DDT with nature is depicted. Note that the pesticide exchanges within physical elements of soil, water, and atmosphere and also within the community of plants and animals, concentrating through a food chain of predators and prey.

Some Chemical Comments

DDT—A Defense of Long Complicated Chemical Names. The name we have been using for DDT is ambiguous. While all of the components of the molecule are included in the name dichlorodiphenyltrichloroethane, the manner in which the parts are assembled is not clear. The correct name of DDT is this jawbreaker: 1,1,1-trichloro-2,2-*bis*-(p-chlorophenyl)ethane. Such a complicated name seems ridiculous on first glance, but it is needed to distinguish DDT from some similar compounds. Some very similar isomers of DDT are shown (pp. 134–5); all could be called dichlorodiphenyltrichloroethane. They differ not in the constituent parts of the molecule but in how the parts are put together. With other such changes, one can generate many more different molecules. All together, dozens of different arrangements of chlorine and phenyls on an ethane structure are possible, and none has the same properties, nor is as active an insecticide as DDT. The longer name we have given for DDT denotes it uniquely among these isomers. The root word of the name *ethane* establishes a chain of two carbon atoms—$H_3C—CH_3$, on which the other structural parts are built. 1,1,1-Trichloro says that there are three chlorine atoms replacing hydrogens on the first carbon atom in the chain—$Cl_3C—CH_3$. 2,2-*bis* says that there are two groups replacing hydrogen on the second carbon and that they are the same; (p-chlorophenyl) establishes that this group is a benzene ring with a chlorine atom bonded across the ring from the site of attachment of the phenyl to the ethane carbon. Since there must be four groups attached to the second carbon, there has to be a hydrogen there also. Thus the name specifically and uniquely describes the structure. Because molecules can differ in small ways, names have to be precise and detailed.

As an exercise it would be of interest to name DDT isomers A to D in Fig. 4.4.

OTHER INSECTICIDES

Once the insecticidal properties of DDT were known, it followed that compounds of similar structure might also be good insecticides. In the years following the development and first use of DDT, several other such compounds were found to have significant insecticidal properties. Many of these first synthetic organic insecticides shared with DDT the structural feature that there were several chlorine atoms in the compound. For this reason they are referred to as the chlorinated hydrocarbons.

Several share with DDT the ethane structure and the presence of phenyl rings (Fig. 4.5).

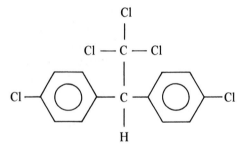

DDT

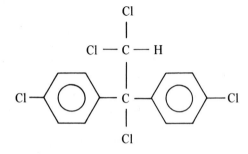

Isomer A

(2 Cl on one ethane carbon, 1 Cl on other)

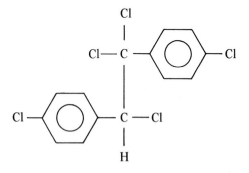

Isomer B

(Phenyl groups on different carbons)

FIG. 4.4 DDT isomers.

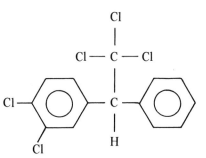

Isomer C

(Both phenyl chlorines on same phenyl)

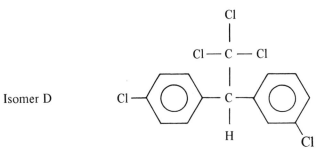

Isomer D

(Phenyl on right has the Cl at meta position)

FIG. 4.4 (contd.) DDT isomers.

Another group of insecticides, developed in the late 1940's, are the chlorinated derivatives of cyclohexane. They are known as *benzene hexachlorides*. This is an example of poorly named compounds since they are structurally related to cyclohexane, not to benzene.

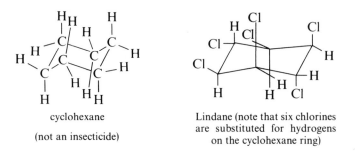

cyclohexane

(not an insecticide)

Lindane (note that six chlorines are substituted for hydrogens on the cyclohexane ring)

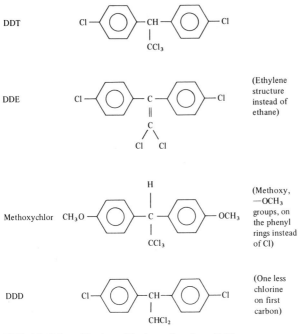

FIG. 4.5 Other chlorinated hydrocarbon insecticides.

Yet another group of chlorinated hydrocarbons are known as the *toxaphenes*. These are cyclic compounds, several rings being fused together.

These two compounds are very similar in structure. Each may be divided into two halves. In one half the carbons are bonded with hydrogen; in the other half the hydrogens have been substituted with chlorine. The compounds differ by the presence of a three-atom ring containing oxygen in dieldrin (on the left sides of the structures as shown in Fig. 4.6), where a double bond is present there in aldrin.

Generally these substances share in common some characteristics with DDT. They are persistent in nature and have become widely dispersed. They also have a high solubility in lipid tissues and concentrate in the food chain. Thus we can discuss them as a group from this point on in our consideration of pesticides.

DIFFICULTIES IN THE BIOSPHERE WITH CHLORINATED HYDROCARBONS

In the course of the quarter century since their use began, some problems have arisen because of these substances as insecticides, and other difficulties

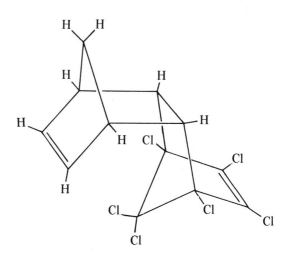

Aldrin

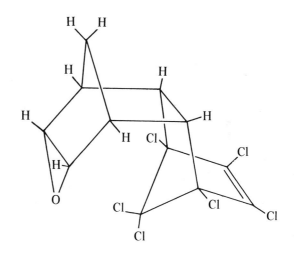

FIG. 4.6 Toxaphene insecticides.

Dieldrin

seem imminent. Among these problems are their accumulation in the fatty tissues of higher animals and their persistence, which we have already mentioned. There are other items of significance. First, insects have demonstrated a developing resistance to chlorinated hydrocarbons. As a result these compounds have proved less effective or have required heavier application as time has gone by. Second, there have been some disturbing developments involving effects of chlorinated hydrocarbons upon the reproductive capacity of some species of birds. Another problem is that natural insect enemies of an insect target are sometimes affected more adversely than the pest. Application of the pesticide may thus have exacerbated the problem, not eased it.

The appearance of resistant species of insects to DDT was noted very shortly after the first usage in the 1940's. Resistance develops because of natural selection, the process that controls evolutionary development. If a few individuals in the insect population have a genetic peculiarity that causes them to have a higher capability for withstanding the pesticide, they will survive and reproduce, while others will not. Their progeny will thus constitute a higher proportion of the next generation, they will withstand the pesticide better and reproduce more, and soon the mutant genetic feature will dominate because the generational turnover in most insect species is very rapid. When that happens, the species will survive an application of pesticide to which they previously would almost all have succumbed.

By 1948, 12 insect species were found resistant to DDT. By 1970, 224 were resistant. The development is probably due to the insects' capacity to synthesize an enzyme* called *DDT-dehydrochlorinase*. This enzyme is named for its biochemical activity, i.e., removing hydrogen chloride from DDT. That causes the formation of DDE as we have already seen, and DDE is less toxic than DDT, so that the insects' resistance is enhanced.

DDE and DDT have been implicated in another serious ecological imbalance. In the 1950's and 1960's several observations by ornithologists in various parts of the world indicated that the populations of some predatory birds were in decline. On closer inspection the reason was found to be that the birds were not reproducing in sufficient number to maintain the population. There could be several reasons for this. One might be that something was causing sterility in the male or the female, or both. Another possibility would be that while normal matings occurred, there was something wrong with the gestation process, so that hatching did not occur. Further along the line, normal hatchings followed by abnormally high fledgling mortality or high mortality among adult birds, so that the number of mating seasons for each pair was reduced, are other possibilities.

Biologists soon showed convincingly that the second possible cause was responsible for the reproductive problems. The specific difficulty was decreasing thickness of the eggshell. The shells were breaking in the course of normal activity in the nest before the embryo was mature enough to hatch. This was especially notable when the thickness of the shells was compared with shells of eggs that had been preserved in museums from the pre-DDT era.

Having found the basic cause of the reproductive difficulty of predatory birds, scientists now had to face a more difficult problem, to find the cause of the reducing shell thickness. An eggshell is made of calcium carbonate, $CaCO_3$. In order to manufacture the shell in the oviduct, the female bird must ingest calcium ion or transport it from other tissues, particularly bone tissues, that are rich in calcium ion. Carbonate ion is readily available from CO_2 in the blood. To produce carbonate requires hydration of CO_2 in the blood, followed by loss of proton.

$$CO_2 + H_2O \rightleftharpoons \underset{\text{carbonic acid—unstable}}{[H_2CO_3]} \xrightarrow{\text{loss of proton}} HCO_3^{\ominus} \xrightarrow{\text{loss of proton}} CO_3^{\ominus 2}$$

This reaction requires the action of an enzyme known as *carbonic anhydrase* to proceed at a fast enough reaction rate. Other enzymes and possibly sex hormones direct the transport of these materials to the reproductive tract. Some change in the synthesis of these enzymes or hormones or some change in their activity could cause the disruption in eggshell synthesis. DDT may be acting in this manner.

Biologists examining affected species such as the California grebe, condor, peregrine falcon, and Bermuda petrel noted that the amounts of DDT and DDE had increased sharply in the tissues of the birds. Predatory birds that feed on fish are in the positions at the end of the food chain and are subject to the enhanced concentration of chlorinated hydrocarbons that we have already described.

It still remains to correlate the two effects. After all, two things that occur at the same time may do so because of coincidence. One is not necessarily the cause of the other. One attempt at correlation deals with thickness of eggshells and DDE content of predatory birds in Alaska. Among several species examined, those with high DDE content also showed the greatest loss of shell thickness. Some species in the same habitat feeding on the same prey showed normal or near-normal shell thickness and low DDE contents (see Fig. 4.7). One might ask why these differences are found since DDE (and therefore DDT) content of the habitat and of the Alaskan food supply would obviously be the same for both species. The birds most affected turn out to be those that are long-distance migrants. One bird, the tundra peregrine, strongly affected in eggshell thickness, winters in Latin America. Another, the Aleutian peregrine, little affected, is a permanent resident of the Aleutians, where pesticide residues are low. The migrators must pass through regions of the North Temperate Zone highly polluted by pesticides, and this appears to be the source of the difficulty. The relation between shell thickness and DDE content of eggs is shown in Fig. 4.7.

Much work remains to be done on this problem. In particular, it is important to understand more about the effect in terms of the specific

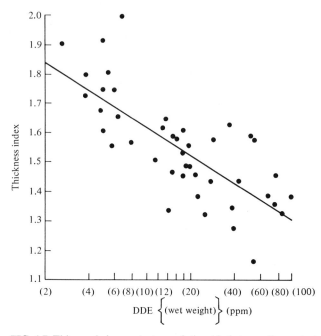

FIG. 4.7 This graph demonstrates a relationship between the content of DDE in the egg and the thickness of the shell, with shell thickness decreasing as the concentration of pesticide is increased. The straight line represents a best fit of all the points which represent individual measurements of eggs studied. It is not surprising to find considerable scatter in these points as with many other measurements of natural phenomena. It also should be pointed out that the correlation is not linear, but related to a logarithmic scale on the horizontal axis. (Reprinted with permission from Tom J. Cade, Jeffrey L. Lincer, Clayton M. White, David G. Roseneau, and L. G. Schwartz, "DDE Residues and Eggshell Changes in Alaskan Falcons and Hawks," *Science*, **172**, 955–957 (1971). Copyright 1971 by the American Association for the Advancement of Science.)

interference of the chlorinated hydrocarbons upon the reproductive process in molecular terms. The other question is how to save predatory birds from extinction. The change to biodegradable pesticides that has been taking place should gradually reduce the DDT and DDE content in birds' food supply, but this is expected to occur slowly and the situation of some of the birds in the early 1970's is so desperate that emergency measures are required. One such approach being used at present is artificial insemination of captured females followed by artificial incubation of the eggs. Some success at this has been reported, and this may save the species until persistent levels of pesticides have decreased.

In 1972 such problems with DDT led to a ban on the use of the compound in the United States, except for rare allowed applications. Interestingly the exportation of DDT is still permitted by the Environmental Protection

Agency, who bears the responsibility for federal pesticide policy. The reason that foreign use of DDT has not been discouraged is that the control of malaria in the tropics has not been particularly successful with other insecticides, and this is a central overriding problem in the underdeveloped world.

Another controversy involving chlorinated hydrocarbons revolves around an insect imported into the southern United States. The fire ant is an example of an imported insect without natural predators in the United States. This insect was introduced into the Mobile Bay, Alabama, area in the 1930's. The species spread over much of the Southeast, although infestation north of the Carolinas has been limited by the colder climate. The fire ant colonies build large mounds. They prey on neither livestock nor crops, but occasionally livestock and humans who disturb the colony suffer very unpleasant stings because the ants swarm furiously to attack an invader. The U.S. Department of Agriculture has embarked upon a program to eradicate the fire ant. Their tactic has been to spray large areas from the air with a chlorinated hydrocarbon insecticide, mirex. In 1969 14 million acres were sprayed. It had been planned to spray 126 million acres altogether at a cost of $200 million.

There have been serious objections to this approach. The indiscriminate spraying technique obviously destroys many beneficial insects, and this

FIG. 4.8 A field in the southeastern United States heavily infested with the imported fire ant. (Photo courtesy of the United States Department of Agriculture.)

approach seems much less efficient than application of the bait directly at the site of the colony, where its application would be more effective against the fire ant and less destructive of nontarget pests. Moreover the pesticide has been shown to persist and build up in food chains similar to DDT. The spraying program has lagged in the 1970's due to opposition by environmentalists and to funding problems.

ALTERNATIVES TO CHLORINATED HYDROCARBONS

The difficulties associated with chlorinated hydrocarbons that we have just described have led those concerned with insect control to seek other kinds of compounds as alternatives. Most important among these have been the organophosphorus insecticides, the most widely used of which are parathion and malathion. Their structures are shown.

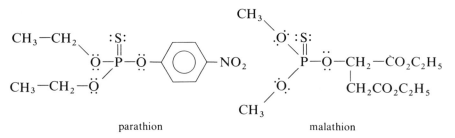

parathion   malathion

Each of these compounds is a derivative of phosphoric acid. We show the structural progress from phosphoric acid to parathion in Fig. 4.9. Such a progression of changes can be useful in seeing the component parts of a structure. Another important insecticide is methyl parathion containing methyl groups instead of ethyl groups.

The chief advantage of the organophosphorus compounds as a group is that they are not persistent. Hence problems associated with chlorinated hydrocarbons—widespread proliferation, buildup in food chains, and concomitant effects in the physiology of plants and animals—are not observed. The chief disadvantage is that the compounds are highly toxic to humans. As a result, great care must be taken by persons applying the insecticides to agricultural crops, and the fields or orchards must be carefully policed to prevent accidental contacts after spraying. Since the pesticides' lifetimes are short, there is no long-term problem. Malathion in particular has a short lifetime and is less toxic. The difficulties with application have led to some loss of life among migrant and other farm workers, adding yet another burden to the others that society forces upon the migrants.

Some problems associated with chlorinated hydrocarbons also exist for organophosphates. The development of resistance to parathion has been

Phosphoric acid: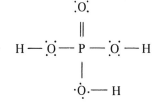

Triethyl phosphate (a triester of phosphoric acid-- an ethyl group is substituted for each hydrogen):

$$CH_3 - CH_3 - \overset{\overset{\displaystyle :O:}{\|}}{\underset{\underset{\displaystyle CH_3 - CH_2 - \cdot\ddot{O}\cdot}{|}}{P}} - \ddot{O} - CH_2 - CH_3$$

The remaining oxygen is replaced with sulfur, which happens to be a member of the same family of the periodic table; this makes the compound a thioester:

$$CH_3 - CH_2 - \overset{\overset{\displaystyle :S:}{\|}}{\underset{\underset{\displaystyle CH_3 - CH_2 - \cdot\ddot{O}\cdot}{|}}{P}} - \ddot{O} - CH_2 - CH_3$$

Instead of one of the ethyl groups in the ester we have the para-nitro phenyl group in the insecticide parathion:

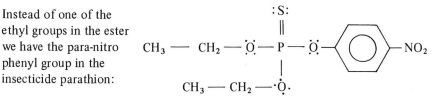

parathion

FIG. 4.9 Structural relationship of parathion as a derivative of phosphoric acid.

found, and the organophosphates are much more destructive of nontarget pests. Thus the natural enemies of the insect being combatted, as well as earthworms, mammals, and birds, will be affected during the period after spraying when the concentration is high.

A third category of insecticides are the carbamates, such as *carbaryl.*

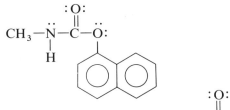

The carbamates may be seen as both esters* ($-\overset{\overset{\displaystyle :O:}{\|}}{C}-\overset{..}{\underset{..}{O}}-R$ group) and amides* ($R-\overset{..}{N}-\overset{\overset{\displaystyle :O:}{\|}}{C}-$ group). They are less toxic to humans than parathion and its analogs, but they are readily degraded to harmless materials, unlike the chlorinated hydrocarbons. The organophosphorus and carbamate compounds act physiologically by inhibiting the action of the enzyme acetylcholinesterase, involved in the transmission of nerve impulses. Since this enzyme occurs in all species with developed nervous systems, one can thereby account for its more general activity.

ALTERNATIVES TO CHEMICAL INSECTICIDES

The problems and potential problems that we have discussed in connection with chemical pesticides have led ecologists, agricultural chemists, and all scientists concerned with protection of the environment and with the preservation of man's health and food supply to examine other possibilities for insect control.

Perhaps the simple destruction of insects is too naive. There are just too many of them, and they can usually replenish their numbers quickly through their rapid reproductive cycles. Are there other possibilities? To seek out such possible approaches, a simple question is in order: What is the point of the insects' greatest vulnerability? The strength and vitality of the insects does not lie in the individuals. They are short-lived and highly vulnerable to weather and to predators. The vulnerability of insects lies both in the fact that they must reproduce rapidly and massively to make up for high mortality rates and in the complexity of insect metamorphosis. The life cycle from the mature reproducing adult to the next generation is frequently only a matter of days. One pair of adults can conceivably be the ancestors of astronomical numbers of progeny through many generations in a single growing season. If reproductive efficiency were reduced, control would be very possible.

The reproductive process has several key events associated with it:

1. There must be a mating.
2. Both male and female must be fertile.
3. The fertilized eggs must hatch, producing immature insects.
4. The immature insects must pass successfully through the immature larval and pupa stages to reach adulthood.

Effective disruption of any of these points can drastically reduce the populations of the insect. In addition there are other ways of controlling the adults. One of the serious difficulties of pesticides would be overcome if the insects could be brought to the pesticides instead of the other way around, for it would eliminate the necessity of dissemination of the pesticides throughout a field or other ecosystem.

One could also control the pest by the introduction of the natural predators of the insect. Insects in their native habitat are normally controlled by all manner of predators that have developed with the insect in the environment. The activity of man has added a new dimension because of his commercial and transportation activities. Insects have been introduced to new areas and have sometimes flourished in the absence of their predators if the food supply situation were favorable. Included among those predators which may be effectively used are those bacteria and virus parasites which can infect an insect.

The Japanese beetle, the elm bark beetle that transmits Dutch elm disease, and the fire ant are good examples of transplanted insects. A major effort of the U.S. Department of Agriculture and U.S. Customs officials is concerned with the exclusion of others whose presence would be a major threat to one crop or another. Among pests of serious concern in this category are the Mediterranean fruit fly and the screwworm fly, which can menace cattle in the southern United States.

The control of the screwworm fly in the United States marks the greatest success story of one approach, sterilization. After World War I it was felt that southern United States would benefit by more diversification in agriculture, and the growth of a beef cattle industry was encouraged. The screwworm fly posed a significant threat. This insect lays its eggs on the cow's hide. The larvae when they hatch eat their way through the hide and feed upon the flesh of the host animal. At the very best the animal is not going to show an efficient rate of weight gain. At worst, a substantial infection of larva will kill him.

To counteract this insect, male screwworm flies irradiated with gamma rays, and thereby rendered sterile, were released in very large numbers. Because their number was larger than the native fertile males, most matings were between sterile males and fertile females. By repeating the release of sterile males for a few months the species was eradicated entirely. In all, 3 billion sterile flies were released in southeastern United States during the program in the early 1960's. It is necessary to maintain a program of vigilance to prevent the reentry of the screwworm fly from Mexico. Occasional small reinfestations occur and the sterilization program may have to be repeated periodically.

Chemical sterilization offers a potential alternative to sterilization by irradiation. It has the advantage that it could be placed in the field and the effects would result in naturally occurring insects. This would be cheaper than sterilization by radiation, but the possibility of sterilizing beneficial insects or other animals, the persistence possibility, and the difficulty of developing resistance all must be considered. Successful field tests on the boll weevil have been achieved because of sterilization of males with the compound apholate, which has an unusual and interesting structure.

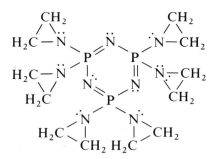

Note that the six-membered ring of nitrogen and phosphorus is similar to the benzene structure.

CHEMICAL COMMUNICATION IN INSECTS

Insects use chemicals for communication purposes, for making trails to food (ants), for signaling alarm, and for attraction of the one sex to the other for copulation. Substances that act in this way as messengers within an insect species are known as *pheromones*. The use of sex attractants to lure the male to the female or vice versa is critical since for many insects such encounters are unlikely by mere chance. The use of chemicals in this way gives an opportunity to disrupt normal reproduction. By the use of attractant in a lure, all of one sex in an area could be lured into a trap, thus preventing their mating with the opposite sex.

A number of pheromones have been discovered in recent years. In the boll weevil, sex attractants secreted by the male attract the female. Four active compounds have been isolated from feces and from extracts of the male insects themselves. The structure of one of these male boll weevil pheromones

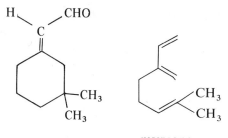

myrcene

FIG. 4.10 Boll weevil sex pheromone.

is shown. It is a member of the class of organic compounds called *terpenes.* Many terpenes are found in plants as aromatic constituents of oils. The boll weevil probably synthesizes the sex attractant from myrcene, shown on the right in Fig. 4.10, which is an ingredient of cotton oil.

The sex pheromone of the fruit tortix is

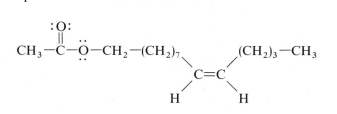

For the almond moth, the attractant is

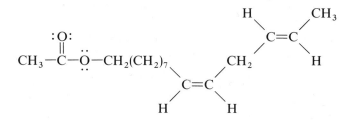

For the pink bollworm moth, another cotton pest, the attractant has the structure

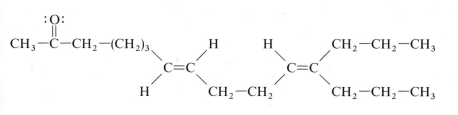

For the gypsy moth, a serious pest of hardwoods in the United States, the sex attractant is a C-18 hydrocarbon containing a single epoxide* ring.

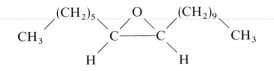

It is noteworthy that as these moth species are related taxonomically, so chemical structures of their sex attractants are in the same chemical family. Yet the structures differ slightly, and the males of one moth species are not attracted to the females of another. This helps prevent any cross-mating within genetically similar species.

Sex attractants show strong possibilities for effectiveness against some of the worst pests, the codling moth, the gypsy moth, and the Mediterranean fruit fly. Their use offers several advantages.

1. The sex attractant is more specific, affecting only one (or a few) species.
2. They can be effectively used with insecticides in lures. This precludes spraying the insecticide on an agricultural crop.
3. Since the reproductive process itself depends on the attractant being present, it is highly unlikely that resistance could develop.

For the chemist the problem has three aspects. One is to determine the chemical structures of the active attractants. Then he must devise a reaction or series of reactions which will lead to synthesis of the material which can be accomplished on a practical basis. Sometimes the possibility of synthesizing compounds of similar structure but even higher activity is possible.

Interesting field tests with attractants have been performed in connection with the gypsy moth. A few years ago DDT spraying against this insect was discontinued in northeastern United States and, as a result, by the early 1970's the insect population had increased sharply causing significant defoliation. Sex attractants used in lures proved effective in preliminary testing in 1971.

One of the striking aspects of the sex attractants is the sensitivity that they arouse. In most insects a dose of the order of 10^{-13} to 10^{-17} g evokes a response in the opposite sex. Typical attractants are active over a range of 100 to 400 ft; in the case of flying insects, sometimes they carry much farther. Because of their advantages, sex and other pheromones should prove useful in insect control in the future.

A few mammals have long been known to communicate through sex attractants. The musk ox is an example. Recent work in biology indicates that many mammals use chemical scents as sex attractants and for other

FIG. 4.11 The male cockroaches are attracted to the glass slide (barely visible beneath them) upon which has been placed a tiny amount of the natural sex attractant produced by the female cockroach. (Photo courtesy of Dr. Martin Jacobsen, Agricultural Research Service, USDA.)

purposes such as establishing territory or marking food. Some research even indicates that sex attractants may operate in primates, the order of mammals that includes man.

Research in the area of chemical communication between living things is a new and exciting field. Plants as well as animals can use chemical communication, and it takes place between species as well as within species. Plants use pollen to give chemical cues to attract pollinating insects. They also can use chemicals toxic to predatory animals. Plant alkaloids (p. 267) are used in this fashion to cause severe toxic effects in mammals, including man. Some plants produce chemicals that act as herbicides thereby eliminating competing plants.

It is clear that plants and animals have developed the capability of synthesis of compounds biologically active on other species as a means both of protecting themselves from enemies and promoting their own welfare. Our understanding of this is infinitesimal, and its understanding will represent a significant advance in biology in the next decades. It will have the result of showing ecology in large measure to be a molecular science.

JUVENILE HORMONES AND INSECT CONTROL

The juvenile hormone approach is an attack on the metamorphosis cycle of the insect. All insects have one or more juvenile stages, the larva and, in some insects, the pupa stages. The changes that take place from the fertilized egg to larva to pupa to the adult are controlled by hormones. One, called the *juvenile hormone*, induces the retention of juvenile characteristics. If harmful insects could be retained at critical points in the larva stage by the presence of excess hormone, no adults would be generated and the reproduction of a new generation of the insect would not occur. The juvenile hormone of *Hyalophora cecropia*, the North American robin moth, has been isolated. Its structure is

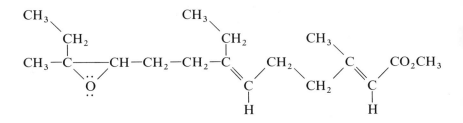

Note the structural characteristics of the compound are a long carbon chain such as the fatty acids, two double bonds, the ester group at the right, and an epoxide ring at the left end as we have drawn it.

Juvenile hormone control of the anopheles mosquito may be the first significant use of this technique. Field studies in the tropics have established significant reduction in populations of the mosquito due to failure of larva to hatch after contact with applied hormone. Anopheles mosquito, the scourge of the tropics because it transmits malaria, has not been effectively controlled by any agent except DDT. This juvenile hormone offers hope of an acceptable alternative to DDT against this pest.

SUMMARY

What is to be done about chemical insecticides and also about the threat of the insects upon man's food supply and his health? Some persons advocate a return to less intensive agriculture using natural methods of insect control and the elimination of chemical pesticides of all kinds. This seems impractical given the food requirements of mankind. The agricultural chemists, insecticide producers, and many farmers advocate application of more pesticides when necessary to effect control, with research and development of new insecticides as the insects develop resistance to those now in use. This theory operates on the premise that the mind of the synthetic organic chemist is more versatile and ingenious than are the potentials of genetic mutations of the insects, probably a correct premise, but one which ignores the complexities of ecosystems.

There is a middle position—that our approach in the recent past has been too one-sided. We relied too heavily on synthetic pesticides and gave too little attention to biological controls, to lures such as the sex attractants, and to the other means we have described.

An example of a more balanced approach is typified by the case of the alfalfa aphid, which can severely damage this important forage crop. The use of chemical sprays can be very effective in conjunction with biological controls when the spraying is properly timed. When applied early in the development of the crop, it can result in sharp reductions in aphid populations without concomitant losses among insect predators. Then the predator has an increased ability to control the aphid, and no more spraying is required. Later spraying damages the predators more severely than the aphids.

Such balanced approaches seem well suited to combat an adversary as versatile, dynamic, and vital as the insects have proved to be.

FUNGICIDES, RODENTICIDES, AND HERBICIDES

Attacks by fungi upon food products and other products made from organic sources are of concern and must be controlled in some manner. Fungi infestation and damage can be costly not only in agriculture and food processing but also in other industries—pulp and paper, textiles, petroleum products, and drugs, to mention a few.

Early fungicides included sulfur or formulations that included salts of the cupric ion. Bordeaux mixture, a mixture of cupric sulfate ($CuSO_4$) and lime (CaO) originally used in the French vineyards, proved successful in many applications. Later organo-mercury compounds, such as phenylmercuric acetate, $C_6H_5Hg^+C_2H_3O_2^-$, were used effectively. Problems associated with mercury toxicity, to be discussed later (p. 176), have since diminished their use.

Another type of fungicide are the dithiocarbamates.

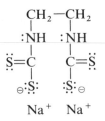

Nabam—disodium ethylenebisdithiocarbamate

Note the similarity to this structure with that of carbaryl, the only difference being that sulfur is in the place of oxygen. The term *dithio* means two sulfur atoms are present.

carbamate group—partial structure found in the insecticide carbaryl

thiocarbamate group—partial structure found in the dithiocarbamate fungicides

Another important type of fungicide includes creosote and other non-volatile derivatives of petroleum. These are used as wood preservatives.

In recent years a new class of compound, called *polychlorinated aromatics*, has been used as wood preservatives. An example is pentachlorophenol.

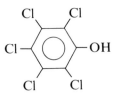

As with the polychlorinated hydrocarbons, problems of residues and lack of biodegradability exist for these compounds.

The damage caused by rats as a destroyer of property and foodstuffs and as a menace to health is measured in the billions of dollars annually. The problem is particularly acute in the urban ghettos and in food storage facilities. An ominous potential of the rat, to act as a secondary host for a flea that harbors the bubonic plague bacillus, is ever present. The principal chemical rodenticide is warfarin. It is structurally related to the compound coumarin.

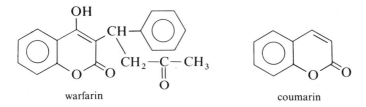

warfarin   coumarin

Warfarin acts as an anticoagulant; that is, it inhibits the clotting of the blood. After an effective dose, an injury, either internal or external, will cause the animal to bleed to death. The bait is usually prepared by soaking grain with warfarin solution followed by drying. Warfarin is active as an anti-coagulant for all mammals, and this leads to the greatest disadvantage associated with its use, occasional accidental ingestion by pets, livestock, or humans. There is some prospect that the future of warfarin as a rodenticide is limited. The reason is that new strains of rats may have been selectively developed against which warfarin is ineffective. Scattered observations of this were reported in 1972.

If dollar value of sales is used as an indicator, the use of herbicides is more widespread in the United States than insecticides. Two substances have been principally used. They are known commonly as 2,4-D and 2,4,5-T.

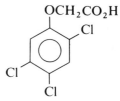

2,4-dichlorophenoxyacetic acid (2,4-D)   2,4,5-trichlorophenoxyacetic acid (2,4,5-T)

Note that the compounds are related to acetic acid and chlorinated phenols in the structure.

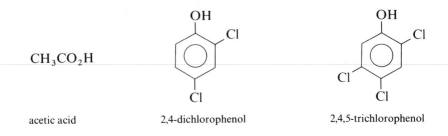

CH₃CO₂H

acetic acid   2,4-dichlorophenol   2,4,5-trichlorophenol

The herbicides act to disrupt normal growth patterns in plants. Their chief advantage is a preferential attack upon broad-leafed plants without harm to grasses or trees. Thus they will selectively kill weeds in a wheat field or in a lawn. 2,4-D and 2,4,5-T offer no problems of residual effects following application, but some questions about birth defects have been raised in connection with 2,4,5-T. When 2,4,5-T was fed to rats, dead and deformed fetuses were produced in statistically significant numbers.

This *teratogenic* effect appears to be due to an impurity in the manufacturing process. The responsible compound is 2,3,7,8-tetrachlorodibenzo-*p*-dioxin.

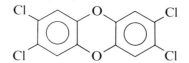

The polychloro derivatives of *p*-dioxin are notoriously toxic. The concentration of this agent in commercial 2,4,5-T is only about 1 ppm. Note that it is formed by condensing two molecules of 2,4,5-trichlorophenol together bridging each to the other through the phenol oxygens and the *ortho* positions. Two molecules of HCl are split out in the process. These results have led to cancellation by the Environmental Protection Agency of 2,4,5-T application on food crops, but its use continues in weed control in rangeland.

Both 2,4-D and 2,4,5-T were used by the United States military in Vietnam for purposes of defoliation of jungle growth, the purpose being to deprive the enemy forces of ground cover. Ecologists have pointed out that such practices lead to drastic changes in the natural equilibrium and that 2,4,5-T could cause health problems in the civilians contacted by the herbicide. Other herbicides were directed against rice paddies, the object being to deprive the enemy of his food supply.

SUGGESTED READING

1. *The Pesticide Review*, published annually by the U.S. Department of Agriculture. U.S. Government Printing Office, Washington, D.C. This document contains pesticide production data and reviews trends in their use.

2. *Cleaning Our Environment: The Chemical Basis for Action.* American Chemical Society Special Issue Sales, 1155 Sixteenth Street, N.W., Washington, D.C., 20036, pp. 193–244. This work contains a comprehensive and critical discussion of pesticides, particularly chlorinated hydrocarbons, entitled "Pesticides in Our Environment."

3. D. E. H. Frear, *Chemistry of the Pesticides,* Van Nostrand Reinhold Company, 450 W. 33rd St., New York, N.Y. 10001, 1955. This is a more advanced discussion of pesticides that is now somewhat out-of-date but contains much basic information.

4. *Organic Pesticides in the Environment,* Advances in Chemistry Series, American Chemical Society Publication, 1155 Sixteenth Street, N.W., Washington, D.C., 1966. This is an extensive discussion of analytical and physiological problems associated with pesticide residues.

5. Rachel Carson, *Silent Spring,* Houghton Mifflin Co., Boston, Mass., 1962. This classic first exposed the pesticide problem. In the decade following its publication, it gained stature more for its prophetic insights into the developing pesticide crisis than for its scientific rigor.

6. *Agriculture and the Quality of our Environment,* Nyle C. Brady, ed., American Association for the Advancement of Science, AAAS Publication 85, Washington, D.C., 1967. This reference contains good summaries of the effects of pesticide residues. These include "Pesticides in Our National Waters" by Richard S. Green, Charles C. Gunnerson, and James J. Lichtenberg; "The Extent and Seriousness of Pesticide Buildup in Soils" by T. J. Sheets; "The Breakdown of Pesticides in Soils" by Martin Alexander.

7. Griffin E. Quinby, Wyland J. Hayes, John F. Armstrong, and W. F. Durnham, "DDT Storage in the U.S. Population," *Journal of the American Medical Association,* **191**(3), 109–113 (1965). This article describes analysis of human fat tissue, taken from subjects in various parts of the United States in 1962, for DDT levels.

8. C. F. Wurster, Jr. and D. B. Wingate, "DDT Residues and Declining Reproduction in the Bermuda Petrel," *Science,* **159**, 979–981 (1968); J. J. Hickey and D. W. Anderson, "Chlorinated Hydrocarbons and Eggshell Changes in Raptorial and Fish-Eating Birds," *Science,* **162**, 271–273 (1968); T. J. Cade, J. L. Lincer, and C. M. White, "DDE Residues and Eggshell Changes in Alaskan Falcons and Hawks," *Science,* **172**, 955–957 (1971). These three references describe specific research studies relating the DDT problem in raptorial birds.

9. Justin Frost, "Earth, Air and Water," *Environment,* **11**(6), 14–33 (1969). A general review of the pesticide residue problem.

10. Deborah Shapley, "Mirex and the Fire Ant: Decline in Fortunes of 'Perfect Pesticide'," *Science,* **172**, 357–359 (1971). This article relates the status of the fire ant problem in 1971.

11. E. Sondheimer and J. B. Simeone, ed., *Chemical Ecology,* Academic Press, New York, N.Y., 1970. A general reference on a new and developing field; M. Beroza, ed., *Chemicals Controlling Insect Behavior,* Academic Press, New York, N.Y., 1970. This latter volume details information on the field of sex attractants to 1970 in several orders of insects.

12. R. H. Whittaker and P. P. Feeny, "Allelochemics: Chemical Interactions between Species," *Science*, **171**, 757–770 (1971). This article is a general reference to this subject.

13. P. M. Poffey, "Herbicides in Vietnam: AAAS Study Finds Widespread Devastation," *Science*, **171**, 43–46 (1971). A report of a study of the use of herbicides by the military in Vietnam.

5

The Metallic Elements
and the Environment

> Man must cease destroying the resources of the planet
> simply because it is cheaper or more efficient to start
> from the beginning than to recover or recycle.
>
> *Max Tishler, President*
> *American Chemical Society, 1972*

Chemical Descriptions of the importance of metals in our culture and
Orientation as a component of modern technology are involved in this
chapter. The reader should review important concepts, such
as the role of oxidation and reduction in chemistry. We shall discuss the
refining of ores to produce metals in terms of oxidation and reduction as they
relate to metals. Sections of the textbook relating to electrochemistry and
electrolysis reactions should be studied. It is also important to review theory
related to the formation of complexes in solution due to coordination of
ligands with central cations. It is advisable to look over sections of your
text concerning electronic structure and configuration as they relate to
properties of metallic elements.

INTRODUCTION

Metals* are exceedingly useful materials because of their high conductivity
of both heat and electricity, their high strength for their weight, the ease with
which they can be shaped, and their durability. Their value as structural

materials is beyond debate. Indeed, the rise of man's civilization is marked by increasing diversity and sophistication in his use of metals. Historians and anthropologists recognize this by using terms like "Age of Bronze" and "The Iron Age" to describe the impact of the development of technologies dealing with mining metals from their ores and incorporating them into the culture as structural materials.

We are going to examine three of the most important metals. We shall be interested in the impact of the mining and refining operations upon the environment. The other problem associated with the heavy use of metals that we shall examine will be the depletion of their natural sources. Later in this chapter we shall examine certain metals that are toxic and pose an environmental threat for that reason.

IRON AND STEEL

The world production of iron and steel is of the order of a half billion tons/year, by far the greatest production of any single metal. About one-fifth of the production takes place in the United States. The industry is considered essential to the economy, perhaps more so than any other commodity. We propose now to examine the iron and steel industry and processes step by step, with special concerns about environmental effects.

The isolation of the metal from naturally occurring minerals or ores is the first step. Iron occurs in most rocks everywhere to the extent of about 5% by weight. The iron occurs primarily combined with oxygen as oxides. The most common of these minerals is hematite, which is the oxide of the ferric* ion—iron(III), Fe_2O_3. Another iron ore is magnetite, which has the molecular formula Fe_3O_4. This formula seems puzzling at first because it means that Fe has a fractional valence state of $2\frac{1}{3}$ if the oxygens are -2. Since this is not possible, another explanation—that the Fe is present in more than one valence state—must be true. Magnetite is an equimolar mixture of the standard oxides of Fe(II)* and Fe(III)*, FeO and Fe_2O_3, respectively.

Enriched iron ore deposits form when igneous rocks (volcanic in origin) are weathered by erosion processes. When this happens, the iron content is increased from 5% or so by weight to higher values as the nonferrous rocks are leached out. A typical ore, taconite, has about 28% Fe by weight. The other materials are silicate rocks and clays. Following removal of the ore from the ground, which is usually done by open-pit operations, the ore is finely crushed and the iron content is concentrated. This is done by forming a slurry of fine particles in water and then running the slurry past a magnet. The iron-rich particles adhere to the magnet and the silicate particles pass over it and are carried away. The fine particles of silicate produced are called *tailings*. The iron-rich particles, now about two-thirds Fe by weight are ready for smelting in the blast furnace.

The largest of these ore concentration operations is carried out near the iron ore sources of the Lake Superior region. The operation is carried out on the Lake Superior shore; the tailings are released into the lake itself, and the enriched ore is transported by ship through the Great Lakes to the blast furnaces of Gary, Pittsburgh, and Youngstown. The taconite tailings in Lake Superior have been the source of much controversy because of the effects on the aquatic ecology caused by slow settling of the particles in the lake. At one site, nearly 66,000 tons are released every day. While the tailings are not thought to contain toxic or nutrient materials, they do form a sediment on the bottom of the lake, and in the region of the release they seriously affect life on the bottom and hence the ecological health of the area. The Environmental Protection Agency has limited the dispersion of the tailings in the lake.

The enriched ore is ready next for the conversion of iron oxides to the metal. This process is accomplished in the blast furnace. It requires the *reduction* of Fe from the $+2$ or $+3$ oxidation state of the iron oxide in the ore to the zero oxidation state of the element. The *reducing agent* is carbon monoxide. It is generated from coke, which is elemental carbon, by oxidation in the presence of a limited supply of oxygen.

$$C + O_2 \longrightarrow CO_2 \qquad \text{produced when oxygen supply is plentiful}$$

$$C + \tfrac{1}{2}O_2 \longrightarrow CO \qquad \text{produced when oxygen supply is limited}$$

Coke, iron ore, and limestone ($CaCO_3$) are injected into the furnace at the top and air is "blasted" in from below. The coke at the bottom is oxidized to CO_2 because the oxygen is plentiful, but as the CO_2 produced rises through the furnace it encounters more coke (elemental carbon), and the O_2 supply is lower. Then the CO_2 and coke react together; the CO_2 is reduced to CO and coke is oxidized to CO.

$$CO_2 + C \longrightarrow 2CO$$

At this point the reduction of iron ore takes place.

$$Fe_2O_3 + 3CO \longrightarrow 2Fe + 3CO_2$$

The iron metal is molten at the elevated temperature of the blast furnace ($\sim 1500°C$). It flows to the bottom and is drained off as crude or *pig* iron.

The function of the limestone is to react with SiO_2 remaining in the ore to form slag. First limestone decomposes in the heat of the blast furnace with loss of CO_2, which reacts as above.

$$CaCO_3 \longrightarrow CaO + CO_2$$

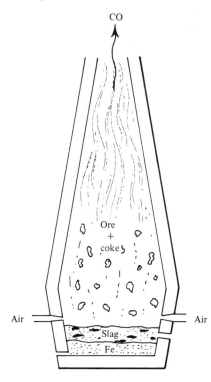

CO

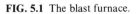

FIG. 5.1 The blast furnace.

Then calcium oxide reacts with silica to form calcium silicate, the principal component of slag.

$$CaO + SiO_2 \longrightarrow CaSiO_3$$

The slag, also molten, flows out the bottom of the blast furnace. It is separated from the more dense iron by gravity since Fe is heavier. The stack gases are conducted from the top of the furnace. This blast furnace process can introduce a number of problems from the environmental standpoint.

1. Sulfur, if present in coke, will convert to SO_2 that will eventually emerge in the gas effluent. Thus coal of low sulfur content must be used in the process to minimize this problem.
2. Carbon monoxide produced in the furnace is a component of the gas effluent and contributes to air pollution. It can be converted to CO_2 by oxidizing it in O_2 as it emerges from the furnace.
3. Slag has limited commercial utility. Much is simply discarded and presents an enormous solid waste disposal problem. Slag heaps are commonplace around some steel mills.

4. Coke ovens pose significant pollution problems. The process of converting coal to coke requires the removal of the organic compounds that occur in coal along with elemental carbon. The conversion is accomplished by heating the coal and in effect distilling off the volatile impurities. The heating is done in the absence of air, naturally, to prevent the combustion of the coal. The volatile distillate is condensed to coal tar, which is a rich source of organic chemicals, particularly aromatic compounds (derivatives of benzene). Coal gas, principally hydrocarbons, is also obtained from the volatile organics. It is used as a fuel within the mill. Emissions from coke ovens include particulates, organic compounds, and SO_2. Most air pollution occurs during charging of the ovens and because of leaks.

Pig iron is not a good structural metal. Its strength is low and impurities are present in large amounts. In particular, a large amount of elemental carbon is picked up from the coke. Sulfur and silicon also are incorporated into pig iron in the blast furnace, and phosphorus and manganese are present too. Steelmaking serves to remove these impurities from pig iron and by the expeditious addition of other metals in small quantities forms metal mixtures or *alloys* with Fe. These enhance or reinforce such characteristics of the steel as its strength or durability. The classical method, the open-hearth furnace, operates to remove the impurities by oxidation, the oxidizing agents being both O_2 and Fe_2O_3 from rusty scrap, which is added with the pig iron into the blast furnace. Fe_2O_3 is used, for example, to remove carbon, while at the same time being reduced to metallic iron.

$$Fe_2O_3 + 3C \longrightarrow 2Fe + 3CO\uparrow$$

Some limestone is used in the open-hearth furnace to react with oxides of the sulfur and phosphorus. This prevents emission of these noxious gases into the atmosphere.

$$CaO + SO_2 \longrightarrow CaSO_3$$
$$6CaO + P_4O_{10} \longrightarrow 2Ca_3(PO_4)_2$$

Silicon and manganese oxides form part of the slag at this stage with the calcium sulfite and calcium phosphate, shown above.

The formation of CO, small amounts of slag, and particulates in the form of ferric oxide are the principal environmental difficulties in the open-hearth operation. The rust-red smoke characteristic of the open-hearth operation is due to Fe_2O_3 particles. Newer processes currently have begun to supplant the open-hearth process for steelmaking. One is the basic oxygen process, in which oxygen rather than air is blown through the molten pig iron-scrap mixture, converting some to Fe_2O_3, which oxidizes the contaminants as before and is in turn reduced back to Fe. Because O_2 is used instead of air, the process is more rapid and hence more economical. Another newer process is the electric furnace in which heat is provided by an electric current through carbon electrodes. The electric furnace has some advantage in being able to handle iron scrap more efficiently.

The most serious pollution problem in the steel industry is the emission of particulates from the open-hearth furnaces. The particulates are Fe_2O_3 and appear as brown-red dust. The particles pose a health problem in areas near the site of emission because particulates can build up in the respiratory system, but the particles soon settle from the atmosphere. With the basic oxygen process the problem remains, but all such furnaces in the United States are equipped with particulate removers. The reduction of Fe_2O_3 particulate emission may be solved in large measure on existing furnaces by the use of trapping devices. The oldest of these is known as the *bag house*—it simply traps some of the particulates on a fibrous "bag." The use of the Venturi scrubber where the particles are washed from the gaseous emission by contact with fine sprays of water and of the electrostatic precipitator are also on the increase. If the particulates are removed from the air by scrubbing, the problem of stream pollution can arise. This can be relieved by allowing water to stand in a clarifying pond, where the particles can settle out before release of the water into a river or lake.

Another difficulty in steelmaking arises in the process known as *pickling*, which is part of the steel hardening process, removing oxide coatings from the metal surface. This requires large quantities of a mineral acid, such as HCl. Before release into the environment, it is necessary either to neutralize the spent acid with an inexpensive base such as lime or to sequester the acid. For example, it is sometimes placed in deep shafts in the ground. The discarding of used lubricating oils from machining, casting, and rolling finished steel operations is responsible for the emission of biochemical oxygen demand noted on p. 98 for the metals process industries.

The most severe environmental problem remains particulate emissions, however. At present, particulate emission from iron and steel industry operations is 13 million tons/year, mostly from open-hearth operations. All basic oxygen furnaces are equipped with control devices. These furnaces now produce about one-third of the steel in the United States and their role is increasing. The replacement of older furnaces with new equipment capable of reducing particulate content in the emissions remains a matter of high priority and concern in the steel industry.

NONFERROUS METALS

The major nonferrous metals of commercial significance and importance in our economy are copper, silver and gold, tin and lead, zinc, cadmium and mercury, and aluminum. To be sure, others play key roles and have important specific and unique functions especially in steelmaking in alloys. With the exception of tin, aluminum, and gold, the metallurgy and environmental problems attendant upon the production of these metals from their ores are fairly similar because all of them occur predominantly in nature as a sulfide of the metal. Moreover, some are found together in the same ore and are isolated in the same or sequential operations. We shall discuss the metallurgy and environmental problems of copper as a typical example.

Copper

Several properties of copper make it as important as it is, perhaps second only to iron among metals. It is an excellent electrical conductor, which makes it important in electrical appliances and transmission lines. Secondly, in contrast to iron, it is fairly resistant to corrosion. Its use in brass and bronze alloys with zinc and tin, respectively, is significant.

Some copper occurs in nature in the metallic state but most consists of cuprous* sulfide, Cu_2S. Note that copper is in the $+1$ oxidation state. This ore is called *chalcocite*. Cu also occurs as a mixed cuprous-ferrous sulfide, $CuFeS_2$, chalcopyrite. The ores contain large amounts of silicates and the first step is to concentrate the copper particles from silicate particles, a process called *beneficiation.* This is done by a flotation method. The ore, finely crushed, is treated with a mixture of oil and water. The copper sulfide ore particles are wetted by the oil; the silicate particles, by water. As the oil floats on top of the water, the copper particles tend to float, and they are removed by blowing air through the oil to form a froth. The silicates sink to the bottom.

The concentrated ore is then subjected to smelting. Air is blown through the mixture and an oxidation-reduction reaction takes place. Sulfur is oxidized from the -2 sulfide state to the $+4$ oxidation state of SO_2. In the early stages of the process, FeS is preferentially reacted when it is present.

$$2FeS + 3O_2 \longrightarrow 2SO_2 + 2FeO$$

Ferrous oxide then reacts with silicon dioxide producing silicates that form a slag along with the naturally occurring silicates. Most of the iron is thus removed.

$$FeO + SiO_2 \longrightarrow FeSiO_3$$

Further roasting produces a crude copper by reducing the cuprous sulfide to the elemental oxidation state.

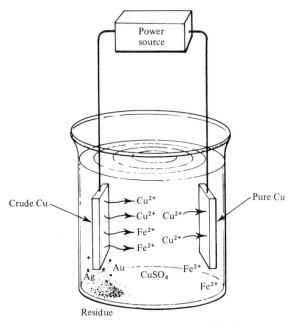

FIG. 5.2 An electrolysis cell for the purification of copper.

$$Cu_2S + O_2 \longrightarrow 2Cu + SO_2$$

Some CuO forms and it also reduces to Cu.

The crude Cu contains some Fe, Ag, and Au as impurities. These impurities reduce the conductivity of the copper and it is necessary to refine further by a process of electrolysis* to obtain a metal of high quality. The crude copper is placed in contact with a solution of cupric sulfate, $CuSO_4$, and connected with a pure copper electrode in an electrical circuit (Fig. 5.2). The *crude* copper is the anode or oxidizing electrode. Cu is oxidized to the Cu(II) ion, which dissolves.

$$Cu \longrightarrow Cu^{2+} + 2e$$

At the *pure* copper cathode or reducing electrode, cupric ions from the solution are reduced by picking up electrons.

$$Cu^{2+} + 2e \longrightarrow Cu$$

Thus the copper atoms transfer from the anode to the cathode, and the concentration of Cu(II) ions in the solution stays the same. The pure copper is recovered from the cathode.

Separation from the other metals happens in the electrolysis because these impurities have different propensities for oxidation and reduction. Iron has a higher oxidation potential,* and so it dissolves readily along with Cu at the anode. But its reduction is more difficult, and so it remains in the solution. Gold and silver have lower oxidation potentials. So, instead of oxidizing, they simply fall unoxidized from the anode and form a sludge at the bottom of the electrolysis cell from which they can be subsequently isolated in other processes, and they become an extremely valuable by-product.

There are two environmental concerns involved in copper production here. For one, the SO_2 produced in roasting in large quantities can cause alarming air pollution. Many smelters are not equipped with controls. About 12% of all SO_2 and SO_3 pollution comes from smelting of ores. The areas near some smelters are strongly affected by high quantities of SO_2 with great harm to vegetation and to buildings, making life difficult, perhaps even hazardous for human beings. The area around some smelters is so denuded of vegetation because of atmospheric SO_2 that it resembles a desert (Fig. 5.3). Fortunately there is a viable solution to the problem;

FIG. 5.3 Sulfur dioxide emissions at high concentration from metal smelters have caused scenes like this one, taken in a national forest in California in 1928. The area has been denuded of vegetation, resulting subsequently in soil erosion. By use of emission controls, many smelters have succeeded in reducing SO_2 levels in the vicinity to more tolerable levels. (Photo courtesy of United States Forest Service.)

unfortunately not as much is done as could be to adopt it. The solution is to remove the SO_2 by scrubbing it from the effluent gas with water, in which it is highly soluble. SO_2 when oxidized produces SO_3, which when added to water yields sulfuric acid, a highly marketable chemical. The expenditures of the nonferrous metal industries for pollution control represent 0.07 % of total revenue for these industries.

The other impact of these processes on the environment is that the use of electrolytic operations places a strain on electrical-generating resources. We shall discuss this aspect of metals technology next since it is an even more important problem for the aluminum industry.

Aluminum

While quite an active metal and therefore easily oxidized in air, aluminum is durable because the aluminum oxide forms a surface over the active metal preventing reaction from proceeding very far. In addition, aluminum has a low density. Because of the low density and resistance to corrosion, aluminum is a very useful metal.

Aluminum is also very abundant, occurring principally as the mineral bauxite, $Al_2O_3 \cdot xH_2O$; x means that some unspecified or unknown number of water molecules are coordinated with aluminium oxide in the ore. Aluminum is refined by reduction in an electrolytic procedure. First, though, bauxite is purified by dissolving it in sodium hydroxide. This is necessary because bauxite is contaminated with iron, which presents problems in the metallurgy, and it must be removed. While most metal oxides, including Al_2O_3, are soluble in acid, forming the metal cation, aluminum is one of the few that also dissolves in base. This usefully separates out the iron that will precipitate in the basic medium as $Fe(OH)_3$ and also as its silicates.

$$Al_2O_3 \cdot xH_2O + 2OH^- \longrightarrow 2AlO_2^- + H_2O + xH_2O$$

Filtration separates the aluminate ion (AlO_2^-) in solution, and treatment with CO_2 acidifies the mixture until $Al(OH)_3$ precipitates. Seeding with crystals accelerates the precipitation. This precipitate is dehydrated to pure Al_2O_3 by heating strongly.

$$2Al(OH)_3 \xrightarrow{\text{1,000°C}} Al_2O_3 + 3H_2O$$

The purified alumina is now ready for the electrolytic reaction. It is dissolved in molten cryolite, Na_3AlF_6, at high temperatures. A carbon electrode is used as the anode. Upon passage of current, molten aluminum appears at the cathode; it is denser than the cryolite solution and sinks to the bottom of the chamber. The cathode reaction is

$$Al^{3+} + 3e^- \longrightarrow Al$$

At the anode, carbon in the electrode is oxidized to CO_2.

$$C + 2O^{2-} \longrightarrow CO_2 + 4e^-$$

From the environmental point of view, the most serious difficulty in aluminum manufacturing is the great power requirement. For the electrolysis, 10 kw-hr (kilowatt hours) are required for each pound of aluminum produced. In fact, aluminum processing requires more energy than any other major utilizer of power. One frequently hears condemnation of the electric toothbrush or other laborsaving devices as the cause of the power shortage problems. The fact is that households are not a major user of power except for space heating and air conditioning. Industrial usage is much more important, and metals processing is the principal user among industries. It would be better to blame the power crisis on the aluminum beer can.

In the United States, the consumption of power by industry represents about 41% of all power consumption. Residential and commercial uses divide the remainder equally. Aluminum production takes 10% of all industrial power. For this reason, aluminum smelters are usually located near hydroelectric power supplies to minimize the cost. As high-grade iron ore deposits are depleted, it is likely that aluminum will be substituted for steel in some uses and this will increase the energy requirements, as seen in Table 5.1.

TABLE 5.1
ENERGY REQUIREMENTS FOR METAL PRODUCTION
(kw-hr/ton)

Steel plate or wire processed from an ore:	2,700
Steel plate or wire processed from scrap:	700
Aluminum from ore:	17,000

THE DEPLETION OF METAL RESOURCES

The needs of civilized man for metals are very great and place heavy demand upon mineral resources. Table 5.2 shows the quantities of some of the major metals extracted from their ores on a worldwide basis in 1 year (1968).

Secondary or recycled contribution is extensively practiced in the United States in various metals processing. Table 5.2 shows the percentages of such major metals that are obtained from scrap. There are considerable problems

TABLE 5.2
STATUS OF MAJOR METALS CONSUMPTION

Metal	World Consumption (metric tons), 1968[1]	Estimated Remaining World Resources (metric tons)[2]	Percent Metal Produced from Recycled Scrap (U.S. only)[3]
Iron	$368,100 \times 10^3$	$129,000 \times 10^6$	31
Copper (refined)	$5,390 \times 10^3$	216×10^6	30
Lead	$3,000 \times 10^3$	48×10^6	46
Tin	185×10^3	$5,000 \times 10^3$	24
Zinc	$5,070 \times 10^3$	65×10^6	7.5
Aluminum	$9,200 \times 10^3$	$1,820 \times 10^6$	21

[1] *United Nations Statistical Yearbook*, 1969.
[2] From Bruce C. Netschert and Hans H. Landsberg, *The Future Supply of the Major Metals*, Resources for the Future, Inc., 1961.
[3] Data is for 1969; Department of the Interior, Bureau of Mines, *Minerals Yearbook*.

associated with recycling metals. For one, the collection and preparation of scrap metal for refuse may be more expensive than extracting the metal from the ores. An industry may limit recycling to keep production costs low enough so that the metal is competitive with other metals or nonmetallic competitors such as paper or plastic products. This is done at a cost to society of an accumulation of solid wastes and in depletion of the mineral resources. In some cases taxes or shipping tariffs discriminate in favor of virgin ore and against scrap. Municipal garbage after incineration typically contains about 30% metal by weight. Possible methods for the recovery of metal from solid wastes are discussed further in the section on solid wastes (p. 188). In the United States, 2.8×10^6 metric tons of iron and 180,000 metric tons of nonferrous metals are included in municipal refuse annually.

Recovery of the metal content of scrapped automobiles is a matter of great interest both for aesthetic and conservational reasons. About 80% of all discarded autos are reclaimed, but the remaining ones either are in auto graveyards or are abandoned by their owners. These latter are definite eyesores scattered about the countryside or on city streets. Those that are reclaimed may be used for scrap steel but the presence of some quantities of nonferrous metals in the body and frame is a serious problem because the steel produced from them will lack the desired properties. Because of the contamination, strength will be diminished, for example.

Scrap-processing research has produced some interesting ideas. One is to reprocess the auto at the taconite ore concentration point of ore processing. The scrap could be shredded into fine pieces and the iron magnetically separated from the nonferrous fragments.

Once an ore has been processed and used, it is no longer available for exploitation. The metal content of the earth's crust is finite, and therefore care is needed to conserve this resource. When a metal precursor begins to be in short supply, there are several possible courses of action.

1. The amount of metal that can be recycled will increase because it is economically more competitive with the ore.
2. Substitutes may be found that can be used instead of the metal in some applications.
3. It may become technologically and economically possible to extract the metal from lower-grade ores or to enhance the efficiency of current extraction and refining processes.

All these options have been and will be used as we face the problem of depletion of metal resources. The difficulties of metal scarcity are bound to be exacerbated by increasing population and by the increasing need of the emerging nations for metal as their technologies and standards of living advance.

As in other cases of resources consumption, the position of the United States is conspicuous by its excess. With 7% of the earth's population, we use about one-third of the mineral ores consumed each year. We are rapidly using up the richest of domestic ores, and in many cases we have been compelled by developing shortages to turn to importing. The United States cannot meet its own needs in several metals including critical ores needed in the manufacture of various grades of steel. These include manganese, chromium, nickel, and cobalt. Tin and aluminum ores are imported. The situations of copper and lead are particularly critical on a worldwide scale (Table 5.2). Unless new discoveries occur, it is probable that more extensive recycling, higher prices, and the substitution of other metals for copper and lead may be required soon.

TOXICITY OF HEAVY METALS

Certain distinctions can be made concerning metallic elements dissolved in our natural waters as cations, and indeed it is possible to draw correlations between the abundance and ubiquity on the one hand with their importance in biological processes on the other. The first distinction is that those cations that are abundant in the marine or freshwater habitats in which biological evolution arose are compatible with life and indeed are required in various processes of metabolism. Because they were there, they were used. Among these are the sodium and potassium cations, Na^+ and K^+, from the alkali metal group of the periodic table; calcium and magnesium ions from among the alkaline earths, and the ferric ion in the transition metals.

A second group are some cations which are present in smaller amounts in natural environments and which are required in trace amounts by the body. Among these are Mn^{2+}, Co^{2+}, Ni^{2+}, Cu^{2+}, Zn^{2+}, Mo^{2+}, and Sr^{2+}. It is characteristic of these metal cations, which are called *trace minerals* by nutritionists, that while their presence in the diet is required, an excess can exert some adverse effects. The case of the cupric ion is typical. This cation is apparently involved in tandem with the ferric ion in some metabolic processes. In the absence or deficiency of Cu, an anemic or iron-deficient condition develops. The minimum requirement is about 2 mg of Cu(II) per day. Too much cupric ion is as bad as too little, however. High dosages of the order of 60 mg/day will produce severe vomiting and diarrhea and other toxicological manifestations.

Finally there are metal cations that are not present in the ecosphere in any but the very smallest concentrations. These cations include Ag^+, Cd^{2+}, Th^{2+}, Pb^{2+}, Hg^{2+}, and Be^{2+}. Among such metals toxic effects can occur at quite low concentrations because human beings and other species have little or no natural defense against them, probably because they did not have to contend with them during their evolutionary development.

The exploitation of mineral resources by man has led to the dispersion of some of these toxic metals into the environment. Instead of being locked in ores in the earth, these metals have been won from their ores, fabricated into useful materials, and then discarded. In some cases toxicological effects have been found among workers exposed to high dosage levels of particles or to solutions containing these toxic metals, some quantity of which was ingested or absorbed. In other instances, toxicity develops after the metal or metal derivative has been chemically changed after release or after long-term effects. In other cases we have only recently realized the potential danger and do not understand its complete implications.

THE ROLE OF METAL IONS IN METABOLISM

Metal cations play at least four key functions in biological systems.

1. Some metal cations serve to maintain an electrical balance in a system, providing ionic strength to blood and other body fluids.
2. Some metal cations can deposit with anions as salts forming hard tissue, such as bones or shells.
3. Metal cations having more than one stable oxidation state can participate in oxidation–reduction reactions.
4. Many metal cations can coordinate with *ligands** to catalyze biochemical reactions.

This final factor may require more explanation.

Some Chemical Comments

The basis for both the nutritional and toxic factors often involves the ability of a metal cation to coordinate with one or more of a chemical species called *ligands*. This interaction is based on the existence of the positive charge on the cation. That charge will prove attractive to anions, and that is why neutral salts made up of cations and anions are stable substances. An attraction can also occur between the cation and an uncharged molecule, if the latter has within its structure areas of negative charge character. Molecules that have one or more pairs of nonbonded electrons are in this category and will frequently also serve as ligands.

The coordination of electron-deficient cations with electron-rich ligands is considered to be an acid-base reaction. The cation is termed a *Lewis acid** (in honor of the American chemist G. N. Lewis) and the ligand is a *Lewis base**.

Among important ligands that one frequently encounters are the following.

1. Halide ions, F^-, Cl^-, Br^-, and I^-.
2. H_2O, or derivatives of water, coordinating through unshared electrons on oxygen. Several water molecules will coordinate with a metal cation in aqueous solution and frequently in a crystal as well. Thus we would be more correct to describe the silver ion in aqueous solution, for example, not as Ag^+, but as $Ag(H_2O)_2^+$.
3. Ammonia or derivatives of ammonia coordinating through the electron pair of the nitrogen atom. In this category are the amino acids* that make up the structure of proteins.*
4. H_2S and its derivatives, coordinating through the unshared electrons on the sulfur atom. These include those amino acids that contain sulfur, such as cysteine.

$$\overset{..}{\underset{..}{H\overset{}{S}}}-CH_2-\underset{\underset{:NH_2}{|}}{CH}-CO_2H$$

5. Acetate ions and other carboxylate anions, coordinating through the oxygen of the anion.

As an example of coordination between Lewis acids and bases, consider the coordination of the silver(I) cation with two molecules of ammonia to produce the complex $Ag(NH_3)_2^+$. First note the electronic configurations:

$$Ag \qquad Kr3d^{10}5s^1$$

$$Ag^+ \qquad Kr3d^{10}$$

The available orbitals for bonding with ligands are the $5s$ and $5p$ orbitals on the silver ion. These hybridize to sp hybrid orbitals which in turn overlap with the sp^3 hybrid orbitals of ammonia in which non-bonded electrons lie, forming two coordination interactions. The reaction is shown in Fig. 5.4. Depending on the orbital situation and the size of the metal cation, two, four, or six ligands will coordinate to the central ion.

The role of metal cations in a biological medium will often involve the coordination of ligands to a cation in a large molecule, or the coordination may occur as a catalytic process in a reaction. One example of the former is in the class of compounds known as *porphyrins*. Chlorophyll is a porphyrin in which a Mg(II) cation coordinates to four nitrogens (see p. 18). Another is heme, p. 174, a key part of the structure of hemoglobin, which transports O_2 through the bloodstreams of animals.

Heme in combination with proteins forms hemoglobin, which transports O_2 from the lungs to cells by coordinating the O_2 molecule as a ligand with Fe(III) cation. The O_2 molecule coordinates in an axis perpendicular to the plane of nitrogen ligands. Myoglobin stores O_2 by the same process in muscle tissue, and cytochrome, which plays a role in the oxidation of organic molecules in cells, also uses iron in this way.

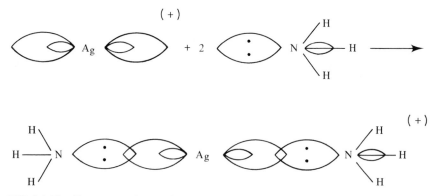

FIG. 5.4 The silver–ammonia complex.

A quite different situation involving a metal cation is presented by the enzyme carboxypeptidase A, which acts to break proteins down into their amino acid constituents. It is secreted by the pancreas into the intestine. The structure of carboxypeptidase A in crystalline form is known and something about its active site has been learned. A Zn^{2+} cation is present at the active site, and it coordinates with the carboxyl group in an amino acid about to be cleaved. The enzyme functions then to break the bond.

It is now clear how a metal cation could be toxic. By substituting itself for the proper ion, the toxic cation could disrupt a biological process, causing it to take place faster or slower than with the proper cation, or it could block an active site altogether.

Several metals have been of concern because of their environmental impact. We shall discuss two at length—Pb and Hg—and three others—Th, Cd, and Be—in less detail.

LEAD—AN OLD KILLER IN NEW GUISE

There is nothing new about lead poisoning. People who have worked in the lead industry or who have otherwise been exposed to the element in the course of their work have been subject to its effects. Lead poisoning results in symptoms mainly in the gastrointestinal tract and in the nervous system. In addition some lead can accumulate in bones. Lead poisoning tends to develop with exposure to small concentrations of lead over a period of time. While the effects are fatal at high enough concentrations or after long enough time, there can be debilitating effects, and a reduction in the general level of vitality due to its action.

The principal toxic form of lead is as the $+2$ cation. Its toxicity arises in its inhibition of the synthesis of porphyrins. Recall that hemoglobin is a porphyrin (see p. 174)—hence Pb(II) acts in effect to lower hemoglobin levels by preventing its replacement. Thus oxygen transport efficiency is impaired. It is possible that at subtoxic levels this could lead to mental retardation or lesser impairment of intellectual capacity. The specific biochemical interference with hemoglobin synthesis is at the stage of the synthesis of the porphyrin center and involves inhibiting an enzyme, δ-aminolevulinic acid dehydrase.

$$CH_3-\underset{\underset{O}{\|}}{C}-CH_2-CH_2-CO_2H$$

levulinic acid

$$\underset{\underset{NH_2}{|}}{CH_2}-\underset{\underset{O}{\|}}{C}-CH_2-CH_2-CO_2H$$

δ-aminolevulinic acid

An intermediate step in heme synthesis is the formation of porphobilinogen by the condensation of two molecules of δ-aminolevulinic acid with loss of two molecules of water.

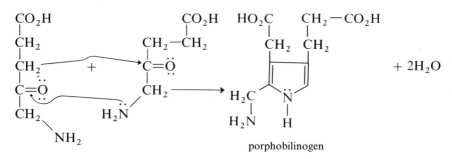

porphobilinogen

This is the reaction in which the enzyme is required. Note that it forms the cyclic structure out of which the porphyrin will be built. It is thus a key step in the synthesis of heme.

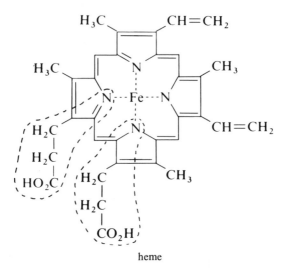

heme

In the structure of heme, the fragments generated from δ-aminolevulinic acid are circled.

There is no line of demarcation in concentration above which constitutes a level of lead high enough to make one sick and below which a person would be reasonably healthy. Some diminution in heme synthesis will do only slight harm; more could do a great deal. The criterion of measurement used is lead ion concentrations in blood measured in micrograms (10^{-6} g) of lead per 100 ml of blood. A dosage of 80 μg/100 ml of blood or higher is taken to represent unquestionably a sickness level, and some think 60 μg/100 ml represents a better lower limit for lead poisoning. It is also generally agreed that persons having 40 to 60 μg/100 ml have sufficiently high levels to be considered in a state of high risk.

It has been established that body concentrations of lead in the population of the United States are increasing. In skeletons of Indians antedating Columbus, the lead concentration has been measured at 4 ppm; in the skeletons of the Civil War era the concentration has been found to be about 40 ppm. In the case of persons alive today, lead levels of 60 ppm and much higher are found *in bones*. The level increases with age also, which indicates that lead is being taken into the body from the environment with time.

Blood levels of lead may be more important in assessing the extent to which a person might be affected, at least for acute poisoning. The mean level in Americans is 20 μg/100 ml of blood. It is possible that subtoxic levels can cause some mental retardation. On the average, mentally defective children have blood lead levels twice as high as normal children.

The ingestion of lead in the digestive tract leads to fairly rapid excretion of the lead in both the feces and the urine. Perhaps only 1 % is retained for any length of time. The buildup requires long periods of ingestion before there are significant effects. Many lead compounds, such as the sulfate and carbonate, are insoluble in gastric juices and hence they are never absorbed into the bloodstream.

The other possible ingestion route is through the lungs. The use of tetraethyllead and similar compounds in gasoline causes some lead compounds, mostly lead chloride and bromide, to be present in the air in particulate form (p. 50). The particles are very fine and may often be described as aerosols. Because of heavy auto concentrations, one frequently finds very high lead levels in air in urban areas especially near a freeway or in a confined parking garage. Lead concentrations in the normal urban air amount to 1 to 3 μg/m^3 of air. Near a freeway during peak traffic hours, levels of the order of 10 to 25 μg/m^3 are common. In a tunnel, 50 μg/m^3 have been measured.

Is atmospheric lead a hazard to health? The question is highly controversial. When air and blood lead concentrations in a high air pollution area,

TABLE 5.3
STUDY OF AIR AND BLOOD LEAD LEVELS IN URBAN AND NONURBAN AREAS

	Average Air Content (μg/m^3)	Average Blood Content (μg/100 ml)
Pasadena, Calif.	4	17.6
Los Alamos, N.M.	0.17	15.4

like Pasadena, California, are compared with the same parameters in a remote desert area like Los Alamos, New Mexico, where auto traffic and pollution are not a problem, the results shown in Table 5.3 were obtained. Note that there is a tremendous difference in lead levels in the air, but the blood levels of the residents are nearly the same. This suggests that atmospheric lead aerosols do not pose a significant health threat. It is known, however, that atmospheric lead can be absorbed into the bloodstream. Thus some of the atmospheric lead probably finds its way into the bloodstream and then is stored in the bones or elsewhere. It should be repeated that this is a matter of great concern and controversy among scientists at present.

Among the 2,255 persons treated for lead poisoning in the city of New York in 1970, all but 90 of them were children 6 years of age or less. Most in fact were under 3 years. How did these children, and only these children, come into contact with lead? There were three clues. Most of the children were found in ghetto areas and other parts of the city with older and substandard housing. Second, until about 1950, many paints contained significant amounts of inorganic lead compounds such as $PbSO_4$, $PbCrO_4$, and $PbCO_3$ used as pigments. When the danger of this was realized, other pigments were substituted in interior paints. The third factor was that toddlers are inveterate chewers and in chewing on painted surfaces, such as window sills, they commonly ingest paint. Even though recent coatings may have covered the lead-base paints, the child would take in all the paints that have been applied since the first, and this would include any pre-1950 coatings.

Many cities in the United States are finding that this is a significant health problem among the urban poor, although no one yet knows how serious or extensive the problem is. It is possible that it may be a factor in learning problems as well. As time goes on and the old housing is replaced, the problem will diminish, although high atmospheric lead content in the cities may counterbalance this. In the meantime stricter housing codes requiring the removal of this paint should be enacted and enforced.

There is another group in the population who are affected by ingestion of lead compounds. These are the drinkers of moonshine whiskey. Many illegal stills have lead in the solder from old auto radiators used as vats and stills. If enough of their product is consumed, adverse health effects due to lead poisoning can be expected.

MERCURY—THE PERFECT PESTICIDE FOR MAN?

Late in 1969 an impoverished New Mexico tenant farmer acquired some grain that had been treated with a fungicide to preserve it. The grain was consumed by hogs who soon showed severe symptoms of disability, especially loss of coordination. Since they were severely ill and of no commercial value, the farmer made a tragic decision—to slaughter one of the animals to feed his hungry family. After a period, three of the children became sick; one went into a coma and two others showed coordination impairment and severe emotional distress.

The tragedy was reminiscent of a still greater disaster at Minamata Bay in Japan in 1955. A large number of residents living near the bay and eating fish taken from the bay were stricken in a similar manner and more than 40 died. In both instances the cause was a compound or compounds of mercury. The grain fed to the hogs had been treated with a fungicide to prevent spoilage. An industrial plant located on Minamata Bay had expelled mercury into the bay. In each case the mercury proved extremely toxic to humans when ingested by them.

These incidents and others have shown that mercury discarded as the metal or as inorganic salts may be converted into other forms that may be toxic to man or other forms of life. In this section we shall look at the conversion process and the threat of mercury toxicity.

Mercury in its metallic elemental state is a liquid at room temperature, and it is thereby unique among the metals. Many of its most important applications have to do with the fact that it combines metallic properties of high conductivity with the liquid properties of mobility. Thus it finds many uses in scientific instruments involving electrical conduction, as contacts and electrodes, and also in temperature and pressure gauges. Most of the other solid metals dissolve readily in mercury forming solutions called *amalgams* that have many applications. In addition to these positive aspects, mercury is not readily oxidized so that it does not degrade with time in contact with air. Mercury is used with silver, to form amalgams that are used in teeth fillings.

Mercury has a small but still appreciable vapor pressure. The metal does not constitute a danger over a short period unless the temperature is increased, which of course will increase its vapor pressure. If a person is exposed to mercury vapor in the air for a long period of time, ingestion of the metal through the lungs becomes a serious hazard.

Ingestion of the *metal* by way of the digestive tract is not a serious hazard. Thus there is little hazard to patients who receive silver amalgam fillings, but some concern has been expressed that dentists and their assistants may run a risk of mercury poisoning because of their constant exposure to small amounts of the vapor. Care must be exerted to minimize spilling of the liquid and to clean up spillage thoroughly. Otherwise the liquid mercury will find its way into crevices in floors and furniture and a constant vapor will be emitted for a very long time. Workers in scientific laboratories must be alert to these same dangers from spilled mercury.

Mercury is produced by roasting cinnabar, the sulfide ore, in a manner analogous to that used for copper (p. 163).

$$HgS + O_2 \longrightarrow Hg + SO_2$$

The most important single use of mercury is in the chlor–alkali process, in which the metal serves as an electrode in the commercial production of chlorine from concentrated sodium chloride solution. In this reaction chloride ion is oxidized at the anode to Cl_2.

$$2Cl^- \longrightarrow Cl_2 + 2e$$

At the cathode, sodium ion is reduced to sodium metal, which dissolves in mercury there to form an amalgam:

$$Na^+ + e \longrightarrow Na$$
$$Na + Hg \longrightarrow Na(Hg)$$

The amalgam is transported into a water solution where sodium reacts to form hydrogen and hydroxide ion.

$$Na(Hg) + 2H_2O \longrightarrow H_2 + 2\overset{-}{O}H + Na^+ + Hg$$

Thus we have three valuable industrial chemicals—hydrogen, sodium hydroxide, and chlorine—from the process. The mercury is recycled back to the cathode where it dissolves more sodium. After a while impurities build up in the mercury so that it must be replaced. The chlor–alkali process consumes 1.5 million lb of mercury per year (Fig. 5.5).

Inorganic mercury compounds are found in two oxidation states. Hg^{2+}, the mercuric ion, is produced by oxidation of mercury with nitric acid.

$$3Hg + 8H_3O^+ + 2NO_3^- \longrightarrow 3Hg^{2+} + 2NO + 12H_2O$$

The mercurous ion is unusual because it exists in all its compounds in the diatomic dipositive state, even though it is in the $1+$ oxidation state. Hg_2^{2+}, the mercurous ion, is produced in a reaction in which Hg is oxidized to Hg_2^{2+} and the mercuric ion is simultaneously reduced to the mercurous

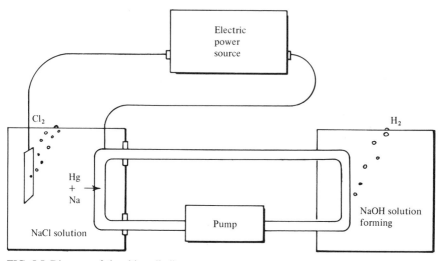

FIG. 5.5 Diagram of the chlor-alkali process.

state:

$$Hg + Hg^{2+} \longrightarrow Hg_2^{2+}$$

The most important salt is the dichloride, known as *calomel*, which is used as an internal medicine and in electrochemical processes. Most mercurous salts are not soluble and if swallowed will not pass into the bloodstream from the digestive tract.

The mercuric ion is perhaps the most directly dangerous form. Mercuric nitrate has long been used to soften fur in the making of felt hats. Because the toxic effect primarily affects the central nervous system, the phrase "mad as a hatter" has been used to describe the afflictions that resulted for workers in the industry. Mercuric sulfate is an important catalyst in some industrial processes. Among these is the production of acetaldehyde from acetylene. The reaction involves the addition of water to the carbon–carbon triple bond followed by a proton transfer that isomerizes the first product to acetaldehyde.

$$H-C{\equiv}C-H + H_2O \xrightarrow{Hg^{2+}} (H_2C{=}C{=}CH) \xrightarrow[\text{transfer}]{\text{proton}} H_3C-\underset{\overset{\|}{:O:}}{C}-H$$
$$H-\overset{..}{O}:$$

The final category of mercury compounds of interest are those organic compounds containing mercury. These may be thought of as alkyl derivatives of the mercuric ion. For instance, if a methyl group with an electron pair forms a bond with Hg^{2+}, the resulting species,

$$H_3C-Hg^+$$

loses one of the positive charges. A dialkyl mercuric compound will be a neutral molecule.

$$CH_3-Hg-CH_3$$

dimethyl mercury

The organomercuric compounds have been known to be deadly poisons for some time. Their principal use has been as fungicides in various industries (p. 151), particularly in paints, pulp and paper products, agriculture, and food processing.

Two facts have recently emerged causing great concern about mercury compounds. One was that, like DDT, organomercury compounds can concentrate in living tissues, so that the concentration in fish, for example, can be much higher than in surrounding waters. The second is that it is possible to convert mercury and mercuric compounds to organic forms in nature. This has been a quite recent discovery, principally due to the work of Professor John M. Wood of the University of Illinois. For years large quantities of mercury have been discarded into water by industrial users, primarily from the chlor-alkali process. It is an ominous fact that this mercury is a potential raw material for the formation of these alkylmercurials.

The toxicity of alkylmercurials is reflected fairly well by the fact that mercury levels in food for human consumption should be below 0.5 ppm. When this concentration is exceeded, the product must be withdrawn from consumption. The margin of safety between levels existing in some seafood and levels that can cause adverse effects in humans is fairly narrow. The threshold for toxic effects is about 0.2 ppm in blood. A reasonable safe limit is probably one-tenth of that. People on high fish content diets may exceed this lower level. Growing children, pregnant women, and persons with nervous disorders may be more susceptible than the average. If anything, an acceptable level of 0.5 ppm of mercury in fish may be too high for comfort. The fact that this is higher than the blood levels and yet considered safe is based on the fact that the uptake from food into the bloodstream dilutes the concentration by increasing the amount of matter in which the mercury compounds are contained. Also, it has not been established that all mercury consumed in food will be absorbed into the bloodstream.

The conversion of the metal or mercuric cation to organomercury compounds, a process called *methylation,* is not yet well understood. There appear to be numerous microorganisms that possess the capability to cause methylation. In one process, a methyl group coordinated to a cobalt 3 + ion is transferred to a mercuric ion. In another the enzyme methionine synthetase is involved. The normal function of that enzyme is to methylate a sulfur, which produces the amino acid methionine. A coenzyme,* called methylcobalamin, is required.

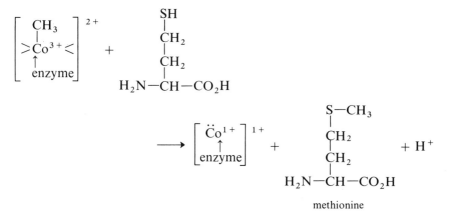

methionine

When Hg^{2+} is present, the same enzyme-coenzyme complex can methylate mercury.

$$\begin{bmatrix} CH_3 \\ >Co^{3+}< \\ \uparrow \\ enzyme \end{bmatrix}^{2+} + Hg^{2+} + H_2O \longrightarrow \begin{bmatrix} H_2O \\ >Co^{3+}< \\ \uparrow \\ enzyme \end{bmatrix}^{3+} + [CH_3Hg]^+$$

methyl mercury

Now that this danger has been established, we need to assess what the future problem is likely to be. It seems certain that too much mercury has already been dispersed and even if no more is released in any of the processes we have discussed, we shall probably face considerable exposure as this mercury is slowly converted to methyl mercury and the other alkyl mercurials. There is at least a remote chance that serious effects for man will develop.

There are several avenues of attack that are open to us, and all are being investigated.

1. We can stop releasing mercury into aquatic ecosystems. This has already been put into effect in large measure. Indeed, since the methylation phenomenon was discovered in 1970, release of mercury has been reduced by 98%.
2. We can recover mercury already released in some way. Much of the elemental mercury from chlorine production plants has been discarded over the years into streams. It was thought that mercury was inert and it would stay in the water indefinitely without affecting the water or life in the water. Since the conversion to alkylmercurials is very slow, much of this mercury still lies in the stream beds and will be converted to the toxic form in time over the years. Recovery is physically possible but it would be extremely expensive.

If we allow the mercury to remain dispersed, we have two options remaining.

3. We can remove mercury from seafood and other sources by some chemical process before it is consumed by humans.
4. We could devise a way of preventing the mercury in tainted seafood from entering the bloodstream from the digestive tract once the food has been eaten. There appears to be at least one hopeful possibility here. It has to do with the affinity of the mercuric ion and its derivative for hydrogen sulfide and its derivatives. Put another way, a very strong complex forms when Hg^{2+} is the central cation and sulfur in the $2-$ oxidation state is the ligand. This is the reason for the toxicity of the mercuric ion in the first place. By coordinating with sulfur groups in amino acids in enzymes and other proteins, it interferes with their normal function. This behavior is in accord with the so-called soft-hard acid-base* theory relative to Lewis acids and bases. Hg^{2+} is a soft Lewis acid and S^{2-} is a soft base; thus they have a high affinity for one another.

Preliminary research indicates that if a polymer containing sulfhydryl groups ($-SH$) is administered to animals along with large amounts of organomercurials or mercuric salts, the mercury is complexed to the polymer through the SH groups. The polymer used involves thiophenol as the basic repeating monomer unit.

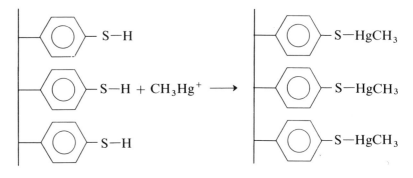

Since the polymer is indigestible if eaten, the mercury is complexed but retained in the intestine. The problem with this approach is that it may well also complex other trace mineral metal cations necessary as nutrients.

A final method of control of methylation of mercury now released into the environment is very indirect, but perhaps the best approach. Since microorganisms are required for the methylation, one could reduce the methylation by minimizing the populations of the methylating microbes. This can be accomplished by reducing the other nutrients in the stream that are the microbes' food supply. Streams in which high methyl-mercury concentrations have been found have also proved to be streams that had large discharges of municipal or industrial nutrient wastes. By reducing the nutrient levels we may reduce the methylating microbe population to the point where methylation is no longer a problem.

This or similar approaches seem hopeful, but we still must learn a lesson from this experience—that before releasing such a material into the environment, we should understand and minimize the potential for ecosystem damage.

BERYLLIUM —A TOXICOLOGICAL MYSTERY —AND OTHER METALLIC CULPRITS

As element four in the periodic table, beryllium is one of the lightest elements. It occurs only rarely however, mainly in the form of its oxide in a mineral known as *beryl*. In other forms, as inorganic salts, or organoberyllium compounds the element has proved highly toxic to humans.

One form of beryllium poisoning has been known for some time. It is an acute toxicity, manifested mainly in the irritation of the skin and respiratory tract, leading to pneumonia-like conditions. It is rather easily diagnosed and treated if the conditions are not advanced. Chronic beryllium disease develops after long-term exposure to concentrations of Be compounds as low as $1 \ \mu g/m^3$ in the air. Symptoms usually appear more than 10 years after exposure, and many organs and systems are usually affected. The disease resembles cancer frequently. That Be compounds were the cause was discovered only after careful and painstaking investigation by physicians in the 1930's and 1940's.

The first cases of the acute form were observed in the 1940's concerning those involved in extracting beryllium from its ores. The first serious cases of the chronic form detected involved workers in the fluorescent tube manufacturing industry. In the early years of fluorescent lighting, beryllium zinc silicate was used as a phosphor. Because of the toxic effects experienced, the use of Be compounds in this role was discontinued in 1949 in the United States. It should be noted in passing that there was a fairly large inventory of unused Be-containing fluorescent bulbs on hand at the time of the ban. These were shipped to overseas customers.

Beryllium metal is used in rocket fuels, and emission from launchings contain some beryllium oxides and halides. Coal has a small quantity of Be and the oxide in a toxic form is emitted when the coal is combusted. The most serious exposures and potential for both chronic and acute forms is localized to workers in beryllium-processing plants, and those who live near such plants. While concentrations of beryllium compounds in the air are usually below levels leading to the acute symptoms, information on toxicity leading to the long-term chronic effects is very sparse. Because diagnosis is very difficult and the problem is of recent understanding, no one is certain how extensive the problem is.

We shall mention two other cations whose toxicity has been of interest. One is thallium(II), which has been used to poison birds of prey such as the golden eagle in the sheep ranges of the Rocky Mountains. A sheep carcass laced with thallium sulfate has been the usual method of administering the poison. It is feared that the use of this salt has so depleted the eagle population that there is the possibility of extinction. Conservationists argue that these birds prey more on coyotes and other sheep predators than on the sheep themselves. A ban on the use of thallium sulfate has been instituted, but recent violations of that law have been revealed. Thallium sulfate is toxic to most other animals including man.

The cadmium ion, Cd^{2+}, has been recently found to be very toxic. Reference to the periodic table will show that it is in the same transition metal family as zinc. Its toxicity arises because of interference with Zn^{2+} in metabolic processes. Cu^{2+} and Fe^{3+}, both important in metabolism, are also disrupted by the action of cadmium.

Presently Cd^{2+} is released at the rate of 15 million lb/year. Principal sources are in coal and petroleum, in paints, from tire tread (which is dispersed on the highways), and from metallurgical processing. Cigarette smoke also contains Cd^{2+} ion.

A little understood source of toxic metals appears to be combustion of coal. We have indicated that it serves as an important source of air pollution of Cd and Be. It is also an important source of mercury pollution in the air. American coals contain mercury varying from 0.2 to 100 ppm. It appears that 90% of this is released as vapor into the atmosphere. The extent to which this may be a health problem in the vicinity of a coal-burning smokestack is unknown.

SUGGESTED READING

1. C. W. Keenan and J. H. Wood, *General College Chemistry*, 4th ed., Harper and Row, New York, N.Y., 1971.

2. H. H. Siser, C. A. VanderWerf, and A. W. Davidson, *College Chemistry*, 3rd ed., The MacMillan Company, New York, N.Y., 1967. These first two references provide good discussions of the metals process industry.

3. D. K. Allen, *Metallurgy, Theory and Practice*, American Technical Society, Chicago, Ill., 1969. This is a more advanced discussion of the subject.

4. Hans H. Landsberg, *National Resources for U.S. Growth: A Look Ahead to the Year 2000*, Resources for the Future, Inc., Washington, D.C., 1964.

5. Brian J. Skinner, *Earth Resources*, Prentice-Hall, Inc., Englewood Cliffs, N.J., 1969.

6. *Resources and Man*, National Academy of Sciences, National Research Council, W. H. Freeman and Company, San Francisco, Calif., 1969. References 4–6 describe the problems of mineral resources, availability, and consumption.

7. Henry C. Bramer, "Pollution Control in the Steel Industry," *Environmental Science and Technology*, **5**(10), 1004–1008 (1971).

8. Three handbooks on the toxicity of metals are of interest: *Industrial Toxicology* by Alice Hamilton and Harriett L. Hardy, Paul B. Hoebner, Inc., New York, N.Y., 2nd ed., 1949; *Toxicology of Industrial Metals* by Ethel Browning, Butterworths, London, 2nd ed., 1969; and *The Chemistry of Industrial Toxicology*, by Harvey B. Elkins, John Wiley & Sons, Inc., New York, N.Y., 2nd ed., 1959.

9. "Mercury: Anatomy of a Pollution Problem," *Chemical and Engineering News*, **49**(27), 22–34 (1971); "Trace Metals: Unknown Unseen Pollution Threat," *ibid.*, **49**(29), 32–34 (1971). These two references review the toxicity of mercury and other metals.

10. "Preliminary Air Pollution Survey of Mercury and Its Compounds: A Literature Review," U.S. Department of Health, Education and Welfare by Public Health Service, Consumer Protection, and Environmental Health Service, National Air Pollution Control Administration, 1969. Other units in this series discuss air pollution associated with arsenic, barium, beryllium, boron, cadmium, chromium, iron, manganese, nickel, selenium, vanadium, and zinc.

11. For a discussion of the roles of metals in metabolic processes, see a standard biochemistry text, e.g., *Biological Chemistry* by H. R. Mahler and E. H. Cordes, Harper and Row, New York, N.Y., 1966. Other worthwhile reviews on the subject are J. C. Bailar, Jr., "Some Coordination Compounds in Biochemistry," *American Scientist*, **59**, 586–592 (1971) and Daryl H. Busch "Metal Ion Control of Chemical Reactions," *Science*, **171**, 241–248 (1971).

12. Daniel T. Magidson, "Half Step Forward," *Environment*, **13**(5), 11–13 (1971). This is a description of problems associated with lead poisoning among children in the St. Louis area.

13. Paul P. Craig and Edward Berlin, "The Air of Poverty," *Environment*, **13**(5), 2–9 (1971). This article depicts the urban lead toxicity problem.

14. M. H. Hyman, "Timetable for Lead," *Environment*, **13**(5), 14–23 (1971). The problem of lead alkyl additives in gasoline is discussed.

15. Robert J. Bazell, "Lead Poisoning: Combatting the Threat from the Air," *Science*, **174**, 574–576 (1971). This is a description of problems associated with high atmospheric lead levels in the New York area.

6

An Untapped Resource —
Solid Waste

> For weeks at a time we saw no sign of man—except his garbage,
> and we saw that all the time.
>
> *Norman Baker*
> *Navigator on the transatlantic voyage*
> *with Thor Heyerdahl from Morocco*
> *to Barbados, 1970*

THE OLD PROBLEM

In February, 1968, a sanitation workers strike occurred in New York and lasted 9 days. As a result about 100,000 tons of garbage piled up on the streets of the city. A health emergency was declared, and the danger to the public safety would have been much worse had not the weather been unseasonably cold. The story is illustrative of yet another pollution problem that has been made worse by the concentration of people into small areas in the urban centers. As the population concentration increases, so does the concentration of all manner of discarded material. While disposal of solid wastes was seldom a problem on the farm, it is one of the most serious problems troubling our cities.

The output of refuse collected from all sources amounts to 360 million tons/year in the United States. This works out to about 10 lb per person per day. Only one-third is from direct household activity. The rest comes from industry, and government operations such as streetcleaning.

The content of the refuse varies wildly from area to area, and season to season, but the breakdown in Table 6.1 is typical. About half of these wastes come into public or private collection agencies through which they must somehow be disposed of.

TABLE 6.1
SOLID WASTE COMPOSITION[1]

Material	Percent by Dry Weight
Paper, including cardboard	46
Food wastes	12
Textiles and wood	10
Plastics including plastic packaging	4
Grass and dirt	10
Glass, ceramics, and stones	10
Metallics	8

[1] *Cleaning Our Environment: The Chemical Basis for Action*, American Chemical Society, Special Issue Sales, 1155 Sixteenth Street, N.W., Washington, D.C. 20036, 1969, p. 167.

Thus about four-fifths of the refuse is of organic origin; the rest is of inorganic origin.

In order to remove this conglomerate from our presence we spend about 3 billion dollars/year. The most common route for disposal of solid wastes is the open dump where about 85% of disposal occurs. There are so many things wrong with the open dump: It is a sanitation problem, runoff from it can contaminate a stream, it attracts vermin, there are odor problems, and the land used becomes unsightly and unsuitable for other use. The only thing good about it is that it is cheap, and this is why it is used by both public and industrial dumpers. Given the magnitude of the garbage needing to be disposed of and the land scarcity in and near our cities, the use of land in this destructive way looks less and less like the right answer.

A vast improvement is the sanitary landfill. In this approach a canyon, ravine, or hilly area is used. Garbage is spread in layers and covered over. As a result the refuse is not exposed. After the area has been completely filled, it may be used for other purposes. Once covered, the organic refuse is decomposed by anaerobic bacteria producing principally CO_2, along with methane and inorganic products. Among the problems of the sanitary landfill approach are the availability of sites and the transportation costs. Care in planning must be exercised in avoiding pollution of groundwater by seepage from the landfill. One of the most successful projects of this kind is being carried out by the Los Angeles County Sanitation District. The use of the sanitary landfill is rapidly increasing across the country.

Another solution is incineration, in which approach the organic refuse is burned in a furnace. The air pollution problem must be controlled, especially since the incinerator is usually in an urban area. The most important pollutant is particulate matter. About 1% of the refuse is emitted in the form of fly ash. In order to control this, electrostatic precipitators are usually required. There is also the problem of disposal of the inorganic ash.

Sometimes ocean dumping offers a viable economic alternative to the open land dump. New York, for example, uses this practice to some degree for solid wastes off Sandy Hook, New Jersey. The practice is widely carried out for dumping dredge wastes from harbors and for toxic industrial chemicals. In the case of solid wastes, dumping provides nutrients for aquatic plant life, and thus eutrophication can develop. Since marine life is concentrated most closely near the coast, refuse would have to be transported to deep water to avoid the problem, and this presents a dilemma, since the cost is thereby increased.

Each of these methods recovers nothing from solid waste refuse. Yet garbage is rich in resources. One can use its nutrients to enrich soil. Recovered metals, glass, plastics, and paper could be recycled. Next we shall explore various methods of refuse treatment that are aimed at recovering some of these components. What are the useful materials that could be recovered? The metals are obviously in this category. The cellulose in paper could be combined with virgin pulp and made into new paper. Glass bottles, of which we discard 26 billion in a year, could be reclaimed and reused or formed into new glass products. Food wastes have value as fertilizers. The problem is that all of these components are mixed together in the garbage bag, in the collection truck, and in the dump or landfill. Thus if reclamation is to be accomplished, the problem is one of separation by either physical or chemical means at a reasonable cost.

SOME NEW SOLUTIONS

The simplest method is separation and recycling of solid waste at the source. The discarder separates refuse into the various categories that can then be recycled. Large producers of wastepaper, such as a publisher or data-processing center, can do this easily. A junkyard is a commercial enterprise dedicated to this function for metals. Rubber tires can be recycled by service stations. Efforts by citizen ecology groups began to concentrate on this approach for glass and paper in the early 1970's by organizing collection stations. In World War II and other times of national stress, recycling by means of paper drives or scrap drives has been used to conserve resources. For a household this separation at the source would require five or six garbage cans instead of one. It would also require attention and energy of the householders that is now directed elsewhere.

There still remains a large quantity of mixed garbage to be dealt with. There are numerous processes that deal directly with mixed solid waste. Some have been in existence for a long time. For instance, composting* has long been used to convert biodegradables into fertilizer by bacterial action. This approach could even be combined with sewage treatment operations. This is done in some western European cities. After separation of bottles, cans, and cardboard, the garbage is combined with sewage sludge and cured until it has attained a fine soily consistency. The problem with composting has been that inorganic fertilizers are cheaper, and such sources of organic enrichment of soils as peat moss or cattle manure are also cheaper.

The need for better systems for recovering and recycling components of mixed municipal solid wastes is the subject of much current research activity. Several experimental systems are in operation. One such has been built as a prototype at Franklin, Ohio, a community of 10,000. Funded by the Environmental Protection Agency, the system separates several components successfully. The first step is to reduce the materials into small pieces. This is done by a process called *hydropulping* (see Fig. 6.1), which reduces the organic components to fine particles which are carried through the process as a slurry in water. Glass and other breakable materials are fragmented into small pieces, of about a 0.75-in. diameter. The metal cans and other large metal fragments, stones, etc., are removed by gravity. The ferrous metals are separated from the others with a magnet. Thus ferrous metal is salvaged readily in good condition.

Glass, dirt, and small metal pieces such as bottle caps are next separated from the organic slurry by taking advantage of their high density. Centrifuging the slurry is done in a "cyclone," and the heavier particles sink to the bottom rapidly and the lighter organic particles do not. The metal content of this reject is high in aluminum and since the density of glass is nearly the same as aluminum (2.7 g/cm^3), separation on the basis of the fact that glass is transparent to light while the metal is not can be done using an optical scanner to identify the fragments for separation.

The organic slurry is next subjected to a screening in which the paper fibers (as well as some rags, grass, and other fibrous organics) are collected on the screen. These fibers can be dried and form a crude pulp that can be used successfully to make tar and roofing papers, low-quality newsprint, and similar grades of paper. One of the most serious problems in the paper reclamation part is the presence of grass and other untreated plant matter, which, as they contain lignin, will reduce the quality of the resultant product. The remaining material in the slurry consists of plastics, food wastes, and other organic substances that were not caught on the fiber screen. By evaporating water the slurry can be converted to sludge and incinerated, composted, or used as fertilizer.

Another process, in Madison, Wisconsin, crushes solid garbage in a dry form, then separates ferrous metals by use of a magnet. The remaining material is buried in a landfill. Other cities, such as Atlanta, Georgia, have similarly extracted ferrous metals from solid wastes by magnetic separation for a number of years. The key test that an experimental recycling system

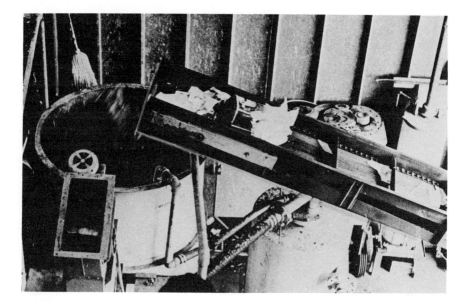

FIG. 6.1 In these photos we see two phases of the Franklin, Ohio, experimental solid waste recycling system. At top, solid waste enters the cycle as it is conveyed into the hydropulper. (Photo courtesy of *Dayton Journal Herald.*) On the left is a later phase of the cycle in which crude cellulosic material suitable for low-grade paper product production is recovered.

TABLE 6.2
ESTIMATED COSTS OF VARIOUS DISPOSAL METHODS,
TRANSPORTATION EXCLUDED

Open dumping	$1.50/ton
Sanitary landfill	4.50/ton
Franklin, Ohio, automated system	6.00/ton
Incineration, including air pollution control	12–14.00/ton

must meet is that of cost. A comparison with the other processes in order is shown in (Table 6.2).

The processing costs in the Franklin plant depend on the amount of refuse handled. As with almost any industry, increasing production levels lower overhead costs. When processing 50 tons of refuse per day, the cost is nearly $40.00/ton, a prohibitive cost. At a 500 tons/day rate, costs significantly decrease, to less than $10.00/ton. Neither of these costs includes such intangibles as the aesthetic improvement, land conservation, and preserving resources that would otherwise be used. Another problem is to "scale up" such an operation. Would a technique that works for Franklin, Ohio, also work for New York?

If stringent budgets that are the general rule in city government prevent widespread use of recycling systems, there are still some first steps that can be taken. Ferrous metal recovery is fairly easy, although the market for scrap has been declining. Another approach is to link up the garbage incinerator with a power-generating facility. The recovery of the energy stored in organic matter by photosynthesis, once recovered, could be utilized by man. Instead of simple incineration, one could use the heat released during the garbage combusion to drive the generators. This would serve to substitute garbage for the fossil fuels and contribute a small saving to that resource.

Research on several other treatment methods shows promise. One is the development of new catalysts that will improve incinerator effluent characteristics. These are employed in catalyst techniques called *fluidized bed reactors* that facilitate combustion. If combustion is more complete, air pollution problems in the incinerator vicinity should decrease. Another research group at the U.S. Bureau of Mines laboratories is working to find methods to reduce the organic components to hydrocarbons. They have had success by heating the refuse under high pressures and have recovered a crude petroleum product.

Such examples show that we have been less than imaginative with our refuse in the past. Some of the alternatives that we have proposed here are viable technologically. They may exact from us a price tag, but wise and efficient use of our resources including garbage will certainly be a more important factor in our way of life in the future.

PACKAGING—A SEPARATE PROBLEM IN SOLID WASTE DISPOSAL

About 50 billion tons of solid wastes from all sources are packaging containers of all kinds that we have used and discarded. The chemical content of packaging materials is not unlike that involved in other refuse, as we see in Table 6.3.

Plastics present a special problem in disposal. Polyethylene (p. 107) is the most common plastic packaging polymer. It has a high degree of stability, which is of course a critical point for a packaging application. Once disposed of, this is no longer an advantage though. While paper and wood can be ingested by bacteria, and thereby degraded, polyethylene is not readily subject to this action. Thus it persists. Attempts to design biodegradable polymers either by including additives or by finding substitutes for presently used plastics are being studied. Another hopeful technique is to introduce additives that would cause the polyethylene to "self-destruct" after its packaging function had been fulfilled.

With polyvinyl chloride a special problem has been found. It is that it undergoes loss of HCl in an incinerator. HCl, when present in incinerator fumes, is a corrosive and unpleasant component.

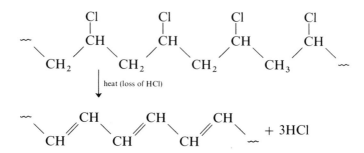

In recent years the consumer packaging market has changed significantly. Plastics and metals have replaced glass, and the disposable container has

TABLE 6.3
PACKAGING MATERIAL USE IN THE UNITED STATES

Material	Percent of Packaging Sales	Percent by Weight of Packaging Material
Paper products	48	38
Metal	25	13
Plastic	11	2.7
Glass	9	45
Wood, textile, et al.	7	1.3

replaced the returnable one. In 1958, 98 % of our soft drinks and 58 % of our beer was sold in returnable bottles. By 1976 the percentage for soft drinks and beer are expected to be reduced to 32 and 20 %, respectively. Even given the costs of handling, cleaning, and refilling it, the returnable bottle is cheaper in both dollars and energy consumption than the throwaway bottle or can. Of course, it should be noted that the marketing success of disposable containers is due in part to consumer acceptance. It is another example of our acceptance of increased conveniences at environmental cost.

SUGGESTED READING

1. *Cleaning Our Environment: The Chemical Basis for Action*, American Chemical Society, Special Issue Sales, 1155 Sixteenth Street, N.W., Washington, D.C. 20036, 1969, pp. 163–193.
2. Charles B. Kenahan, "Solid Waste: Resources Out of Place," *Environmental Science and Technology*, **5**(7), 594–600 (1971).

These first two references describe in more detail some of the developments in solid waste technology treated in this chapter.

3. Thomas D. Clark, *Economic Realities of Reclaiming Natural Resources in Solid Waste*, U.S. Environmental Protection Agency, 1971. This reference details the cost problem in recycling.
4. Michael Heylin, "Consumer Packaging," *Chemical and Engineering News*, **49**(16), 20–23 (1971).
5. Bruce M. Hannan "Bottles, Cans, Energy," *Environment*, **14**(2), 11–21 (1972). This article uses computer techniques to analyze the total social and environmental impact of disposable and returnable soft drink containers.

7

The Chemical Detectives: Pollution and Analytical Chemistry

> The greatest single contributor
> to environmental pollution is ignorance.
>
> *From* Man's Impact on the Global Environment,
> *Report of the Study of Critical Environmental Problems,*
> *M.I.T. Press, p. 150*

Chemical Orientation The reader should refer first to those sections of his text relating to the discipline of analytical chemistry to acquaint himself with its philosophy and general methods. In dealing with pollution problems in this chapter, several key analytical methods are used and review of key sections of your text covering these methods is in order. Among these methods are extraction, techniques of chromatography, and spectroscopic methods. It is also important to review concepts of chemical equilibrium and electronic structure of atoms in connection with the chapter.

Is there appreciable DDT present in the tissue of a marine organism? Is a power plant producing significant amounts of particulate emissions? Is the BOD downstream from a paper mill high enough to cause ecological problems? Is the concentration of lead in the bloodstream of a ghetto child in St. Louis at a dangerous level? Is it possible to identify the tanker responsible for an oil slick?

Before government can take action against a polluter, or before an industry can assess the extent of its cleanup problem, it is necessary to determine the facts. This function lies largely in the province of analytical chemistry with respect to environmental pollution. The role of the analytical chemist is to use his own ingenuity in conjunction with modern instruments and with the knowledge and methods that have been developed through the history of chemistry—to obtain the facts.

The problem is frequently difficult because only a very small quantity of the pollutant is sometimes sufficient to cause a serious pollution effect. Thus the methods must be very sensitive. Moreover it is necessary to determine the concentrations of the compounds being analyzed, not just to verify their presence. In this section we shall look at the role of analytical chemistry in three pollution problems—pesticides in the environment, air pollution monitoring, and analysis for metal cations in solution.

ANALYSIS OF PESTICIDES

In our discussion of DDT and other chlorinated hydrocarbons, we mentioned problems associated with residues in soil, water, and living tissues, including human beings (p. 132). Since the pesticides are typically present in a few parts per million or even less, the problem is not routine. One may begin by separating DDT and the other insecticides from the major constituents of the sample. After this separation, a reliable and efficient analytical method must be used.

The removal of DDT from any of the sources above may be accomplished in a fairly straightforward manner, by taking advantage of the lack of polar character in the DDT molecule (see p. 130). For this reason, DDT will have a high affinity for a solvent of low polarity such as a hydrocarbon solvent, in preference to an aqueous phase, including most living tissues. We can design a chemical experiment that can separate out the insecticide residue by taking advantage of this property.

The technique is called *extraction*. It works by setting up a competition between two immiscible liquids that are in contact with one another. In the case of pesticide residues, one could use a hydrocarbon solvent like hexane with water. Since they are mutually insoluble, two layers will form when they are combined, with the hexane, being of lower density, on top of the water layer. Every solute component will be distributed between hexane and water according to the relative affinity of each component for water. Since DDT and the other chlorinated hydrocarbons much prefer hexane, they will be found in that layer almost entirely. The distribution between the two layers may be expressed as an equilibrium as shown.

$$\text{DDT (hexane)} \rightleftharpoons \text{DDT (water)}$$

The relative affinity may be expressed in terms of an equilibrium constant.*

$$K = \frac{[\text{DDT (water)}]}{[\text{DDT (hexane)}]}$$

In our example, K will be very small since the denominator concentration will be large and the numerator concentration will be small.

Most other compounds present except pesticides will be soluble in water, in preference to hexane. That includes the inorganic salts and polar organic compounds, such as sugars. Polymeric materials like cellulose or proteins will be soluble in neither solvent and would be removed by filtration prior to the extraction. The same would be true of clay and siliceous materials in soils. A direct extraction with hexane on the solid or semisolid material is a slight variation on a two-solvent extraction that could be used to isolate chlorinated hydrocarbon insecticide residues. Separation of the two layers in an extraction funnel is simple (Fig. 7.1). One allows the water layer to flow out the bottom while the hexane layer is retained in the flask by shutting off flow with a stopcock at the line of demarcation between the layers.

The hexane layer is dried with a chemical drying agent to remove traces of water. Such a drying agent is anhydrous sodium sulfate, which picks out the water molecules and incorporates them in its own structure by forming a hydrate. Next the hexane solution is reduced in volume by evaporating some of the solvent. This increases the concentration of the solute, which is not

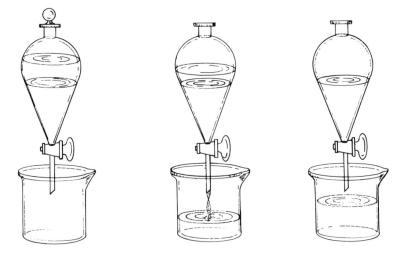

FIG. 7.1 The technique of extraction is illustrated: left, two solvents in contact; center, lower solvent is removed through the stopcock of the extraction funnel; right, stopcock is closed retaining the upper solvent, which is then removed through the top of the funnel.

evaporated, to a more workable level, of the order of 1 % of the solution by weight.

In the next operation a technique of analysis known as *gas chromatography* is used to detect specific pesticides in the solution. Typically at this point we could have a number of pesticides present with many naturally occurring materials also. Gas chromatography serves to distinguish all of these pesticides both qualitatively and quantitatively from one another and from other compounds.

Chromatography is very like extraction, but with an added twist. Like extraction, chromatography is based on distributing a compound or mixture of compounds between two discrete phases. The phases may be two immiscible liquids, a gas and a liquid or a solid and a liquid. The twist is that one of the phases moves with respect to the other and, in so doing, the possibilities for useful separations are even greater than for extraction.

One form of the technique is called *paper chromatography*, in which the stationary phase is a strip of paper or, more precisely, the water present in the fibers. The mobile phase is a solvent system that is immiscible with water. This technique is used to analyze amino acids. The mobile phase is brought into contact with a strip of paper at one end. The moving solvent works its way by capillary action upward through the paper and comes into contact with the sample. As it moves further, it will separate the sample components depending on their relative affinity for the paper and the moving solvent. *Column chromatography* uses a solid with high surface activity as the stationary phase and a solvent that runs through the solid in a column and is eluted out as the mobile phase. In a cross between paper and column chromatography, the solid may be spread into a thin layer through which the solvent moves by capillary action.

The technique of *gas chromatography* uses a gas as the moving phase and a nonvolatile liquid imbedded on an inert solid base as the stationary phase. To explain gas chromatography and chromatographic effects in general, let us set up a typical experiment. As solid support we shall use finely divided firebrick onto which has been adsorbed a standard high-boiling liquid phase. This material is packed in a column in a piece of stainless steel tubing, which is formed into a coil and placed in an oven in the gas chromatograph instrument, where the temperature may be controlled. We adjust the temperature as required and pump helium as the mobile phase carrier gas through the column.

Now we begin the experiment by injecting a sample of our concentrated residue into the chromatography column at a temperature high enough to volatilize it completely in the flowing helium. Only a few microliters (millionths of liters) of the 1 % solution are required because the technique is sensitive. This is its principal advantage, because the chemist must frequently analyze material when he has very little with which to work. The sample once

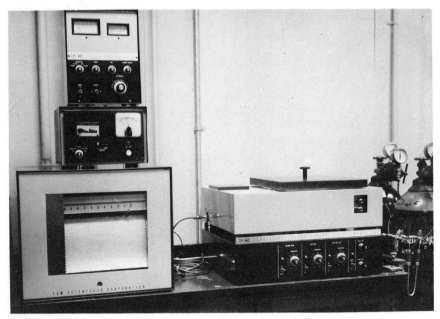

FIG. 7.2 A standard gas chromatograph unit. The carrier gas helium is supplied by the tank on the right, the injection port and oven area is in the center, and the recorder is on the left with electronic controls for the detector and oven temperature at the top left. (Photo courtesy of Monsanto Chemical Company, Research and Engineering Division, Dayton, Ohio.)

injected into the flowing gas stream will be in position to interact with the liquid phase. This interaction is the key to the separation by gas chromatography.

Suppose for the sake of simplicity that we take a two-component pesticide residue containing aldrin and DDT as our sample. Suppose further that DDT has a higher affinity than aldrin for the liquid phase we are using. If we were to look at the system with an imaginary stop-action camera, we would find that each of the two pesticides sets up an independent equilibrium between adsorption on the liquid and remaining in the vapor phase.

$$DDT \text{ (adsorbed)} \rightleftharpoons DDT \text{ (vapor)}$$

$$aldrin \text{ (adsorbed)} \rightleftharpoons aldrin \text{ (vapor)}$$

When we say that DDT has a greater affinity for the liquid phase, we mean that K_{DDT} is less than K_{aldrin} in these equilibria. If an equimolar mixture of the two is injected, the liquid composition will be richer in DDT than aldrin and the vapor composition will be richer in aldrin as the chromatography begins [Fig. 7.3, stop action (1)].

Now recall that the vapor is moving through the column. If we move the vapor in this equilibrium to a new part of the column, we have the new stop-action situation (2). In the new section of column [stop action (3)], some of each compound will condense, but again the vapor will be richer in aldrin and the condensate in DDT. In the earlier sections some of the condensate will vaporize into fresh carrier gas with the same tendency for aldrin to vaporize more than DDT occurring again. As the mixture proceeds along the column, it is possible for us to predict a result. The two components will begin to separate from one another [stop action (4)] with aldrin preceding DDT through the column. If the column is long enough and the difference in affinities for the liquid phase sufficient, we shall arrive at the situation in stop action (5). The column has done its work. Aldrin will be eluted from the column first and DDT later, both in pure form.

After the column, next comes a detector of some kind. Two types are common: the *thermal conductivity detector* and the *flame ionization detector*. By measuring the heat conductivity of the effluent gas stream and comparing it to a reference stream of helium that has not gone through the column, the

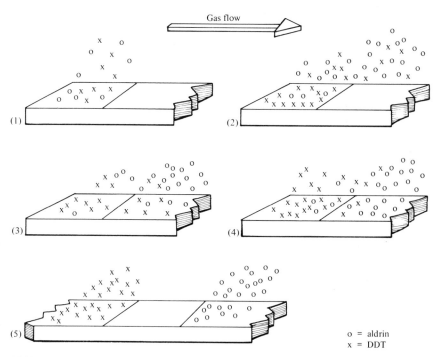

FIGURE 7.3

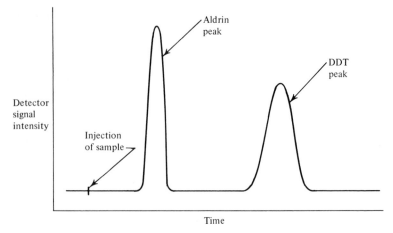

FIG. 7.4 Gas chromatogram of DDT–aldrin mixture.

thermal conductivity device can detect when a sample component is emerging by the fact that the thermal characteristics of the effluent gas are then different from the reference. An electrical signal is generated and sent to a recorder. The flame ionization detector operates by converting the molecules emerging from the column into ions after combustion. These ions generate a current and this current is recorded.

From either detector a chromatograph is produced, a recording of the current as a function of time. When a small quantity of a substance is eluting, a small signal is generated. A large amount gives a large signal. When nothing is emerging, there is no signal; the recorder is on the base line.

By recording the signal on a moving chart, a permanent record can be produced (Fig. 7.4). A peak usually shows a normal statistical distribution as shown in the graph; that is, the peak begins with a small amount of material eluting. More and more compound elutes from the column, generating a stronger and stronger current through a maximum. Then the signal decreases as it is increased. There is a tendency toward peak broadening the longer the substance stays on the column. That is why the DDT signal is broader than the aldrin signal in the chromatogram.

Quantitative analysis is possible because the peak intensity is a function of how much material is present. One must take into account differences that arise from differences in conductivity; but when this is done, the quantity of each component can be determined from the area under the peak, which can be measured simply.

The problem of qualitative analysis, identifying the substance generating each peak, is a tricky one. If the substance is known, one can run it through the same column under the same conditions. If the substance is the same, it will appear after the same length of time. This period is called the *retention time*. It depends on the temperature, the flow rate of the gas, and the stationary phase as well as the compound itself. Identical retention times do not constitute proof of identity, though. It could be coincidence. To assure identity, the effluent substance can be collected coming off the column and subjected to further analysis, even though only perhaps 1 μg can be collected. One example is to analyze by mass spectrometry. This technique causes fragmentation of the molecules in a manner unique and characteristic for a given structure in the same way that fingerprints are unique for persons; that is, every substance gives a distinctive and unique *mass spectrum*. By interfacing gas chromatography and mass spectrometry, an unambiguous qualitative and quantitative analysis of a separable mixture is often possible.

The technique can of course be used to include any number of components in a mixture, although peak overlap occurs now and then. Gas chromatography is especially valuable because it is useful for measuring very small quantities of materials that are volatile, or at least volatile at high temperatures. These two characteristics are typical of problems associated with pesticide residues.

ANALYSIS OF TRACE METALS BY SPECTROSCOPIC ANALYSIS

The techniques most useful for detecting the presence of small amounts of mercury and/or other metals or their cations is to measure the frequency and intensity of absorption of light in the visible and ultraviolet regions of the spectrum of a sample. This approach can be used because the absorption of energy by atoms, molecules, and ions is selective and discontinuous (p. 18).

Some Chemical Comments

Whether we are talking about an atom, a molecule, or an ion, we know that certain discrete energy levels or orbitals exist. In the normal situation the electrons in a given species will take up residence in those orbitals of lowest energy, each orbital being occupied by two electrons of opposite spin.

As we described in Chapter 1, electrons may absorb energy and be promoted to orbitals of higher energy to form excited states. Electronic transitions* of this kind in atoms and molecules usually take place at frequencies occurring in the ultraviolet* and visible* regions of the spectrum. Since there are several occupied and unoccupied orbitals in a system, there will be a number of different absorptions occurring.

It will be possible for us to determine where the absorptions take place by shining light of different wavelengths in sequence upon a substance or solution and then determining those frequencies which are absorbed and those which are not. The technique is known as *absorption spectroscopy.*

The first problem is to generate light that is monochromatic, that is, in which all photons have the same wavelength or frequency. Normal white light is a mixture of many different wavelengths. It is possible to generate monochromatic light by use of a prism, which spreads light in a rainbow. By turning the prism, we can measure what happens—absorption or no absorption—by stationing a detector on the far side of the sample. The method is shown in Fig. 7.5.

If we plot absorption versus wavelength, we shall find transmission of light except at frequencies appropriate to the energies of transition in the sample.

Since the orbitals are different for each element or substance, the ΔE values are different and hence the spectra are different. In fact, for a given

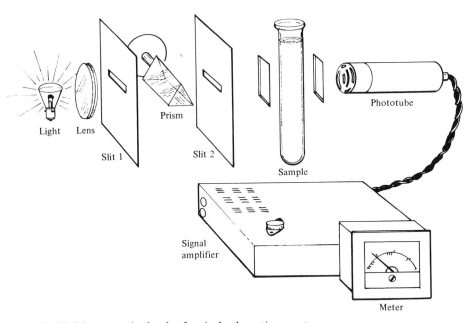

FIG. 7.5 Diagrammatic sketch of a single absorption spectroscopy apparatus. Light generated is resolved by the prism into monochromatic light that can be sequentially exposed to the sample by turning the prism. The absorption of light by the sample is detected by the phototube which generates a signal which is then amplified and recorded.

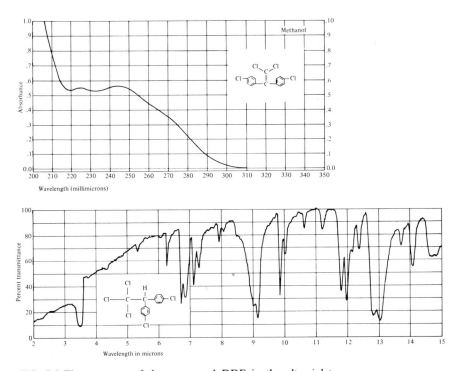

FIG. 7.6 The spectrum of the compound DDE in the ultraviolet region is shown in methanol solution. Two absorption maxima are visible, occurring at 245 millimicrons and 225 millimicrons. The absorbance increases sharply below 220 millimicrons because of absorption by the solvent. The spectrum at bottom is the infrared absorption for DDT. In this case, the wavelengths are of longer frequencies in the micron region, and we are plotting transmittance rather than absorbance along the vertical axis. You will note that the spectrum contains more detail than the corresponding ultraviolet spectrum, and this is frequently helpful in determining the identity of an unknown compound. (© Sadtler Research Laboratories Inc.)

substance, the spectrum is unique and characteristic of the substance and can be used like fingerprints to identify it.

Atomic Absorption—Spectroscopy in Analysis of Metals

In distinction to the broad absorption spectrum shown for DDT, the spectrum for a metal in the vapor phase will show sharp lines indicating the absorption or emission of energy by the individual atoms. The reason for this difference is that the spectrum of a molecule is broadened by different vibrational levels that are possible for the ground state and each excited state.

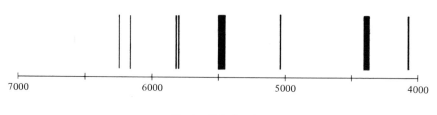

Wavelength in Ångstroms

FIG. 7.7 Atomic emission spectrum for mercury.

For mercury vapor, there are no vibrational levels because there are no bonds. Hence the sharp emission line spectrum shown in Fig. 7.7 is obtained, with nine absorption bands shown in the spectrum.

In addition to qualitative analysis, spectroscopy may also be used for determination of the quantity of a substance. This may be done by measuring the intensity of the light absorbed compared with a sample of known concentration. To determine mercury concentrations, the absorbance of the band appearing at 253.7 nanometers in the ultraviolet region is measured in tandem with a mercury-vapor lamp emitting at that frequency. The absorption of energy is measured against a blank standard. The technique may be used to measure mercury compounds in both the air and in solution.

For nonvolatile salts, atomic absorption spectroscopy is a useful technique for dealing with concentrations as low as 0.1 to 1.0 ppm in solution. To

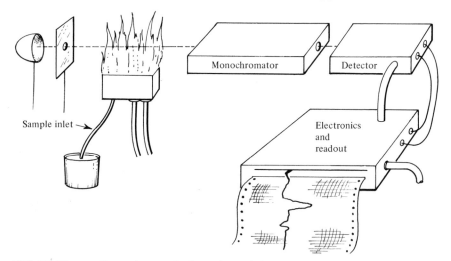

FIG. 7.8 Diagram illustrating atomic absorption technique. Atoms of the metal to be analyzed, after introduction to the flame, will absorb radiation emitted from the same metal present in the source. The absorption is detected and recorded as in Fig. 7.5.

analyze a sample for mercuric salts, a mercury-vapor lamp is employed as the source. The sample is introduced in the form of an aerosol into a flame produced by burning acetylene or a similar gas with O_2. In the intense heat, 1500–3000°C, the water evaporates quickly and the ionic salt will dissociate in part to its neutral atoms. The mercury atoms will absorb radiation at 253.7 nanometers thus decreasing the intensity of the radiation striking the detector on the other side of the flame. The absorbance is thus related quantitatively to the concentration of mercuric salt in the sample solution.

To analyze for lead, a discharge tube with a lead cathode is used instead of a mercury-vapor lamp. The lead cathode emits radiation at wavelengths characteristic of lead, and from the flame ionization technique the absorption of lead from the unknown may be measured. To analyze for other trace metals with the technique requires only that the cathode of the appropriate metal be used.

AIR POLLUTION ANALYSIS AND MONITORING

Suppose that you are the air pollution control administrator for a municipality of 2 million people. You are charged with enforcing areawide municipal air pollution control regulations that empower you to bring offenders to court. Your city and environs are heavily industrialized. You have a typical array of power plants, steel mills, paper mills, and many small industries. All of these contribute to air pollution. Your inner city is tied to its suburbs by freeways that are inadequate during rush hours. Generally there is enough wind to drive off air pollution, but during the summer a thermal inversion occurs occasionally and this causes increases in pollutant concentrations to levels that can cause discomfort or even illness to some people. Your first problem is one of analytical chemistry. Before enforcement of laws against offenders, it must be determined that a violation has occurred, that the output of pollutants exceeds allowable concentration limits. Before a hazard to public safety can be established, a determination of concentration of harmful pollutants must be made. Moreover, the techniques used must be sufficiently sensitive to detect levels of pollutants occurring in very small quantities. The problem is compounded by the very large number of compounds that must be analyzed. As we have seen, typical urban air will contain fly ash from incinerators and fossil fuel-burning power plants, other particulates such as iron and lead oxides, and such gases as sulfur and nitrogen oxides and carbon monoxide, to name only the most prevalent.

Obviously, the analytical problem is challenging. In the analysis of particulates, for example, three different objectives are required. We need to know the amounts of particulates produced, the distribution of particle sizes, and the chemical composition of particulates. These three parameters will vary greatly with location of sampling devices, weather conditions, the time of day, and (of course) the pollution sources.

Typical sample collectors collect particles either on a filter as they settle or by drawing air through the filter and collecting particles. The filter is then weighed, and after making correction for weight changes in the paper, the difference in the weight gives the weight of the particulates. In a cyclone collector a filter is fitted over a drum beyond which a blower pulls air through the filter. After collecting for a period of 24 hr and knowing the capacity of the blower, we can determine the weight of particles collected per unit volume of air drawn through. This level is typically in the range of micrograms of particulates per cubic meter of air.

Determination of particle sizes can be done by optical methods. A stream of air containing suspended particles is passed across a light beam, and the particle size is determined by reflectance of the beam as a function of the particle diameter. For a large number of particles it is necessary to use computer data processing to arrive at a final distribution of particle sizes. The chemical composition can be established by standard techniques such as atomic absorption spectroscopy. Different techniques will be required for different compounds.

Where one chooses to sample particulates is very critical. A station upwind from a power plant will not detect appreciable emissions, while another directly downwind will record values much higher than the typical air in the neighborhood. Usually several stations are located in a city and at different heights above ground level. In order to obtain a fair assessment of the general air conditions, another approach is to use a mobile sampler located in a truck.

For another problem, let us take the automobile exhaust situation. In the effluences from a car one will find again a variety of pollutants. The compositions will depend on the fuel used and the age and condition of the engine, whether it is idling or running and whether there is a pollution control device in the exhaust system. Depending on these factors, particulates, nitrogen oxides, carbon monoxide, hydrocarbons, products of incomplete oxidation such as aldehydes and ketones, olefins, and perhaps lead compounds will be present. The most desirable analytical approach is to probe the flow of gas from the exhaust and analyze continuously for as many components as possible.

To analyze for CO, NO_x, and hydrocarbons, one can insert a probe into the exhaust and measure by infrared spectrometry. This technique measures bond vibrations* in molecules. Vibrational transitions are like electronic transition, in that they absorb specific energies. Thus to promote a vibrating bond to a higher level of vibration is a discrete quantized phenomenon. Only specific energies will cause the transitions, and these energies will differ

FIG.7.9 A high-volume particulate collector. The filter is seen in the center, and air is drawn through the filter, in which the particulate matter is collected. (Photo courtesy of Research Appliance Company.)

for different bonds or molecules. Most vibrations occur with excitation in the infrared region at lower frequencies and lower energies than electronic transitions.

By measuring the amount of energy absorbed for a certain specific wavelength, one can determine the concentration in the effluent of the specific pollutants. An infrared spectrophotometer can determine this by measuring absorption at a specific wavelength against a standard composition. To measure hydrocarbons, for example, one can measure the intensity of the vibrational excitation of the carbon–hydrogen bond at a wavelength of 3.0 μ (microns).* The absorption of the carbon–oxygen bond at 4.6 μ is used to measure carbon monoxide content. Gas chromatography and flame ionization frequencies can also be used for hydrocarbons in auto exhaust.

Technological advances now permit the use of mobile analysis units for auto emissions. Installation of units in service stations would enable protective maintenance to be performed. New autos produced that meet acceptable emission level standards may produce more emission after being driven many miles, or when the engine needs tuning. Preventive maintenance could be used to maintain an acceptable emission standard if changes in performance could be monitored regularly as a standard part of servicing the car. The principal barrier at present is the cost per unit to a service station owner. The costs for CO and hydrocarbon analysis units is $3,000 to $5,000. This is not too expensive if there is enough demand, and this demand could readily be supplied by laws requiring maintenance of acceptable auto emission standards through the life of the car.

The application of analytical chemistry to air pollution is more extensive when one considers the needs for monitoring. The situation is complicated by the numerous compounds to be analyzed, variations of climate, time of day, and location. In order to attain a profile of an air pollution control district, or an urban area, all this information needs to be processed by computers before the situation can be properly assessed, and rational and effective controls can be implemented.

SUGGESTED READING

The following references may be used for further reading in the areas covered in this chapter.

A. Analysis of Pesticides

1. D. J. Lisk, "The Analysis of Pesticides Residues: New Problems and Methods," *Science*, **170**, 589 (1970).

2. "Determination of Organic Insecticides in Water by Electron Capture Gas Chromatography," by W. L. Lamar, *et al.*, in *Organic Pesticides in the Environment*, Robert F. Gould, ed., American Chemical Society Publications, 1966, pp. 187–200.

3. Gunter Zweig, ed., "Analytical Methods for Pesticides, Plant Growth Regulators and Food Additives," **1**, *Principles, Methods and General Applications*, Academic Press, New York, N.Y., 1963.

B. Analysis of Trace Metals

4. G. D. Christian and F. J. Feldman, *Atomic Absorption Spectroscopy, Applications in Agriculture, Biology and Medicine*, John Wiley & Sons, Inc., 1970.

5. J. D. Brooks and W. E. Welfram, "Trace Mercury Determination," *American Laboratory*, **3**(12), 54–57 (1971).

C. Air Pollution Analysis and Monitoring

6. *Air Pollution Control Guidebook for Management*, A. T. Rossano, Jr., ed., Environmental Science Service Division, 750 Summer Street, Stamford, Conn., 1969.

7. W. Lodding, *Gas Effluent Analysis*, Marcel Dekker, Inc., New York, N.Y., 1967.

8. *Air Quality Criteria Documents on Particulate Matter, Carbon Monoxide, and Hydrocarbons.* See Reference 11, p. 62. U.S. Department of Health, Education and Welfare, March, 1970.

9. Werner Strauss, ed., *Air Pollution Control*, John Wiley & Sons, Inc., New York, N.Y., 1971.

8

Pollution and Politics

Decisions·such as the fate of DDT
are not decisions solely within the purview of the scientist
for him to make in the solitude of his laboratory.
Rather they are basic societal decisions about what kind of a
life people want and about what risks they are willing to accept to achieve it.

William D. Ruckleshaus
First appointed Administrator
Environmental Protection Agency
in an address to the
American Chemical Society, 1971

The surprising thing about pollution is that everyone accepted it for so long. Industry tended to take the view that theirs was the right to expel anything for which there was no economic return into waterways or air. Communities with polluting industries took the view that the benefits to the community outweighed the unpleasant backlash of pollution. People said "That's the smell of the payroll" or "That's the price of progress." With little local concern, government at all levels was insensitive to the problem, neither moving for enactment or enforcement of pollution abatement nor putting its own house in order. It seemed that everyone looked upon pollution as an unfortunate but nonetheless acceptable side effect of industrialization. We owe much to those few individuals and organizations who pointed out these problems loudly and often enough that all of us came to appreciate the problem.

With the public awareness and concern now manifest, we can look toward solutions. Obviously corporations and all other polluters must stop releasing noxious or toxic or nutrient materials or at least reduce the amounts released. When this is done, the problem will be solved. We must keep in mind that if pollutants must be recycled or destroyed chemically before release or if operating procedures must be more rigorously controlled, there will be a cost. Who shall pay this cost? That is a political question. We can require industry to bear it as a part of their production costs. If nothing else changes, this will mean lower profits to the shareholders of the corporation. If competition and the law permit, however, the corporation will generally prefer to pass the cost to its consumers rather than its stockholders. Government may choose subsidization of the industry in the form of tax benefits or direct grants as preferable on the theory that the whole public benefits from pollution abatement. In any case, the important point is simply that in one form or another the cost is going to be borne by large numbers of people, whether as stockholders, consumers, or taxpayers. Indeed, the cost will fall on everyone. The realization of this is most important, for only with public support can government continue to take necessary steps in dealing with environmental problems.

There is a school of thought that blames the problem on corporation greed or other failings of capitalism. While it is true that some in industry will fight change in a capitalistic economy, it is also true that pollution is an equally serious problem in such socialist nations as Sweden and the Soviet Union. The philosophy of producing goods cheaply, if necessary at the expense of the environment, must be changed in either a free or managed economy. In the end, the problem remains that it is the consumption of the products of industry that is the problem, and it places upon each individual the moral obligation to make right the environmental damage that his consumption causes.

The arbiter in the decisions required to clean up existing environmental pollution is government at all levels. It is the agency that can referee and resolve the sometimes conflicting advocacy for jobs and profits on the one hand with environmental quality on the other. At the federal level, pollution control is not a new development. There have been several agencies working in the field since early in this century. It is the intensity of concern for environmental quality that is new. In the late 1960's, as increasing concern focused upon the pollution question, more agencies became active, and more money was allocated. Still the agencies were in different departments and there was competition and overlapping. In 1970, President Nixon ordered the formation of a single agency, the United States Environmental Protection Agency, which brought most federal programs and agencies into one unit. The EPA was charged with mounting a coordinated attack in the areas of air and water pollution; solid waste management; and pesticide, noise, and radiation control.

Of all its responsibilities, a cardinal one is that it is the agency that sets and enforces standards within established law. Working in accord with the states when needed, the EPA determines what levels of pollutants are harmful and may establish timetables for the removal of a pollutant. The best known examples of this authority have been the ban of the use of DDT in the United States beginning in 1972, except for a few isolated instances (p. 140). The EPA requirement that auto emissions of carbon monoxide, hydrocarbons, and nitrogen oxides be reduced to 90% of the 1970 model emission levels by the 1975 model year was challenged by most of the auto industry (p. 60), causing a delay until 1976 in implementation.

The second important function of the EPA is research and monitoring. Research is carried out at three major research centers: Research Triangle Park, North Carolina; Cincinnati, Ohio; and Corvallis, Oregon. There are also other laboratories and field stations throughout the nation. The Corvallis laboratory concerns itself primarily with ecological research. Activities at Cincinnati center on pollution control techniques, and the North Carolina center is concerned with research in the field of health effects of pollutants. In addition to its own research, EPA funds research at universities through grants and contracts. It is also necessary for EPA to monitor pollution levels through the use of analytical chemical methods such as those described in Chapter 7.

EPA also serves as an information center providing technical assistance to state and municipal agencies with environmental problems and also to the general public. It also plays a major role in educating environmental experts. Funding for major projects such as the construction of 12,000 municipal sewage treatment plants between 1971 and 1975 has been handled by EPA.

The basis of EPA action is founded in the law, and hence a major federal responsibility lies with the Congress. A number of important items of environmental legislation have been adopted, especially since the mid-1960's. This discussion will touch upon some of the major ones. The Rivers and Harbors Act of 1899 requires that the discharge of substances into a river or harbor may not take place until a permit has been secured from the Army Corps of Engineers. Until 1966 this provision had been taken only to limit large debris that might pose a threat to navigation, but a court decision at that time held that it applied as well to water effluents.

The important water quality laws were enacted in 1956, 1965, and 1970. These laws provide for states to set water control standards for interstate and coastal waters by 1972. The standards vary depending on the classification of the stream use. Most streams are classed for recreation or fish and wildlife propagation uses. Once standards are established, the law provides that both the states and EPA take enforcement responsibility. Violation notices may be given, followed if necessary by prosecution. Sometimes formal conferences between the EPA and state agencies are held to outline enforcement actions. EPA action of this sort played a major role in effecting significant reduction of mercury discharges in 1970 (p. 181). Under the Water Pollution Control Act of 1970, oil spills are prohibited and fines are provided for offenders. Similarly, discharges of sewage from ships into navigable waters is controlled.

The federal entry into the area of air pollution legislation began in 1963. The Clean Air Act of that year provided for assistance to states, the establishment of air quality criteria for pollutants, and a federal role in interstate abatement activities. After added amendments in 1965 and 1967, the Clean Air Act of 1970 placed the federal government in the position of establishing national air quality standards and provided for state implementation. The activity is largely centered through air quality regions based on common geography, degree of urbanization, and other factors. The 1970 act also established the enabling legislation for the auto emission standards we have described.

In the solid waste field emphasis is on research programs to find new waste disposal technology. It is under this program that the Franklin, Ohio, waste treatment plant has been funded (p. 189). The EPA has also embarked upon a campaign to reduce the numbers of existing open dumps, now about 10,000.

The Federal Insecticide, Fungicide, and Rodenticide Act of 1947 and the Federal Food, Drugs, and Cosmetic Act of 1938 along with amendments to those acts, form the basis for federal pesticide control as exercised by EPA. They require registration and EPA approval with the producer being required to prove that the product is both effective and harmless to other animals and to humans. Even after approval, EPA may suspend registration and thereby ban the product. The agency also provides levels for pesticide residue in foods and for research in the field.

Federal authority on radiation lies jointly with EPA and the Atomic Energy Commission. This applies in particular to monitoring nuclear power plant secondary coolant water. Radiation effects on human health is another EPA research area. In 1970, the Noise Abatement and Control Act gave EPA the authority to execute research in noise effects on health and the extent of the problem.

An important new feature of federal planning both within EPA and all other agencies is the Federal Environmental Impact Statement. This document requires any agency planning a project of any sort to submit an impact statement that analyzes probable effects upon the environment of the new program. Whether it be a dam, a freeway, or a new military aircraft, it must be justified on environmental grounds.

It is important to note that much of the pollution abatement program must lie either with the states directly or through the water or air quality regions. Intrastate waterways and such activities as strip mining will largely be controlled by state agencies under the EPA guidelines. There is also the international aspect. To Americans the most direct problem with an international basis is the water pollution of the Great Lakes that is being handled jointly by Canadian–United States consultation. There are also worldwide problems such as the increase in atmospheric CO_2 levels or the tanker traffic and potential for oil spills on the high seas. Handling such problems is more difficult because of the different political jurisdictions involved. A first attempt was the United Nations Conference on the Environment held in Stockholm in June, 1972.

The role of the chemist is important in both the research and monitoring. In pollution monitoring the analytical chemist is concerned with measurements of pollutants and with data processing. Moreover, he is charged with the responsibility of devising more sensitive techniques of measuring lower concentrations and in more sophisticated data-processing systems. Environmental research chemists of every type will be involved frequently with scientists from other disciplines in wide-ranging programs.

SUGGESTED READING

1. "Public Policy and the Environment," A symposium presented Sept. 9, 1969, at a national American Chemical Society meeting in New York, N.Y., the papers in the symposium are published in *Chemical and Engineering News*, **48**(6), 74–81, Feb. 9, 1970.

2. Leo Marx, "American Institutions and Ecological Ideals," *Science*, **170**, 445–982 (1970). An analysis of the American "system" and its ability or inability to respond to the environmental crisis.

3. Readers who wish more specific information on Congressional legislation should consult *United States Statutes at Large*, U.S. Government Printing Office, Washington, D.C., for the appropriate Congress and session.

PART III

THE ENVIRONMENT WITHIN

9

The Population Problem

So God created man in his own image,
in the image of God he created him; male and female
he created them. And God blessed them, and
God said to them, "Be fruitful and multiply,
and fill the earth and subdue it."

Genesis 1 : 27, 28

Or what man of you, if his son asks him for a loaf,
will give him a stone? Or if he asks for a fish,
will give him a serpent?

Matthew 7: 9, 10

Chemical Orientation   In connection with our study of population problems, we shall treat the subject of the sex hormones. You should review discussions of the chemistry and biochemistry of steroids, the structural class to which sex hormones belong. You may wish to explore the wider class of natural products known as *lipids*,* of which both steroids and prostaglandins, another hormone class, are members.

"MULTIPLY AND SUBDUE"

"You're not going to do anything about the pollution problem until something is done about the population explosion." This is one of the battle cries of the environmental movement. The contention behind this statement is that each new human being places a stress on the environment. He needs food, shelter, transportation, and so on. Disposal of his sewage wastes and solid wastes are required and he consumes energy. It may be possible to reduce the amount of these materials required, but it is not possible to eliminate them entirely.

Population is not the only factor in the environmental crisis. Differences exist between the per capita consumption of energy and resources in developed and underdeveloped parts of the world (p. 14). The United States uses more than 50 times as much energy per capita as India because of advanced technology. In either situation, however, an increase in the population will increase the resource demand and hence the environmental stress. Thus the concern for population control is justified and we need to examine it. We shall give attention to the involvement of chemists in the development of technology to limit the birth rate. Before we examine that question, there is need to obtain an overview of the problem.

The first question we need to answer is "What is the extent of the present population crisis and why has it happened?" In the broadest context the answer to the latter lies in the success of mankind in competing with other species for the earth's resources. No other species has had such success. The reason for it is the enormous capability of the human brain compared to other animals. Man has flourished because of his ability to adapt to and even to control his environment. There are many important contributory factors to the rise of man; among them are developing agriculture, learning to use materials, controlling his microbial enemies, and developing complex and interdependent social structures. The rise of modern science and technology has played a major role in the advances that characterize the period since the Renaissance.

Such a successfully adapting species is subject to certain biological benefits. It is bound to flourish. Man has been able to reduce his death rate drastically by increasing his food supply and by improving shelter, sanitary, and medical practices. But as with any successful species, as the number increases, the demand on resources to maintain the population becomes more intense. The tendency is for overcrowding of the habitat and for straining its food supply. These factors serve to limit the population by increasing the death rate again.

The increasing population through history tells the story quite clearly. At the beginning of recorded history there were an estimated 5 million human beings on the earth. By 1650 the population was a hundredfold greater, 500 million. The next 500 million were added in 200 years, by 1850. To 1930 was the time required for increase to a second billion. The third billion arrived by 1960, and the fourth by 1975. These figures indicate another complication. The more people you have, the faster they produce more people. This doubling time, given present rates of growth, is 37 years, while in the past it was very much longer, as seen in Table 9.1.

The growth rate of population is similar to other natural phenomena familiar to the chemist such as the rate of decay of radioactive species or the rates of some chemical reactions. In each case the rate of the process depends on how many units or potential reactants are present. If we wish to describe

TABLE 9.1
WORLD POPULATION AND REQUIRED DOUBLING TIME[1]

Date	Estimated World Population	Time for Population to Double (years)
8000 B.C.	5 million	1,500
1650 A.D.	500 million	200
1850 A.D.	1,000 million (1 billion)	80
1930 A.D.	2,000 million (2 billion)	45
1975 A.D.	4,000 million (4 billion)	37

[1] Paul R. Ehrlich and Anne H. Ehrlich, *Population, Resources, Environment: Issues in Human Ecology*, W. H. Freeman and Company, San Francisco, Calif., 1970, p. 6.

the population at some future time N_t, we can calculate it from the growth rate r, the time span t, and the present population N_0, according to the equation

$$N_t = N_0 e^{rt},$$

where e is the natural logarithm.*

If we have a 2% growth rate per year, $r = 0.02$. For a present population N_0 of approximately 4 billion and a period t of 30 years to about the end of the century, we could calculate N_t.

$$N_t = 4 \times 10^9 e^{0.02 \times 30}$$

$$\frac{N_t}{4 \times 10^9} = e^{0.60}$$

Taking the natural logarithm and solving mathematically gives

$$N_t = 8.6 \times 10^9$$

or a population of 8.6 billion people. This calculation tells us that, with a growth rate of 2%/year, the world population will be twice as great at the turn of the century as it is now. Since this is about the current percentage worldwide, we can expect this population to be generated unless something happens to decrease the growth rate. Moreover, the same growth rate in the twenty-first century would produce a population of 64 billion people in the year 2100.

Opinions among population experts differ about the maximum population that could be comfortably and adequately sustained, but 64 billion is near the highest of such estimates, and many believe the optimum population is much less than that. Irrespective of this argument, it is clear that the natural resources of the earth would quickly be exhausted at current per capita consumption levels, especially if we expect to maintain the standard of living of the United States and other developed countries and to bring under-developed nations to that level. Life is likely to be far more difficult and grim if population pressures even approach this rate. When the maximum level is reached, the natural limits will take command. The population will level off due to an increase in the death rate, a decrease in the birth rate, or both.

To many it seems desirable or even imperative to limit population at a level far below this unknown maximum. Indeed some insist that even present levels are too high, especially if the underdeveloped areas are to approach the standards of living and resource consumption of the developed nations. The "zero-population growth" movement holds this view. Given that the limitation of population is desirable at some level, the next question is one of technique. There are some methods that could be used but that all of us consider brutal or savage. They have to do with increasing the death rate. Clearly we rule out infanticide or extermination of the old, the weak, or the unproductive. Indeed we are still doing everything that we can to continue to decrease the death rate. Having brought most communicable microbe-caused diseases under control, we now seek to conquer cancer and to reduce death from cardiovascular diseases and to reduce the debilitating effects of malnutrition. If the death rate increase possibility is not open, we then must turn our attention to the reduction of the birth rate.

There is some evidence on the basis of the western European experience that suggests that the death rate decrease coupled with industrial develop-ment leads to a natural diminution in the birth rate as the population increases. For instance, consider Sweden. Data from 3 years suggest this has occurred in the last 2 centuries as the death rate has been brought down (Table 9.2). Since the data for 1955 precedes use of oral contraceptives on a massive scale, the decrease in birth rate must be attributed to some other effect. Similar trends are seen in population statistics for other European countries and for the United States. Even so, the rate of population growth remains about 1 % for western Europe.

TABLE 9.2
BIRTH AND DEATH RATES FOR SWEDEN[1]

Year	Birth Rate per Thousand Population	Death Rate per Thousand Population
1785	33.85	27.81
1885	20.00	18.94
1955	14.81	9.65

[1] N. Keyfitz and W. Floeger, *World Population—An Analysis of Vital Data*, University of Chicago Press, Chicago, Ill., 1968.

TABLE 9.3
BIRTH AND DEATH RATES IN 1969
BY GEOGRAPHICAL AREA[1]

Continent	Birth Rate (1969) per Thousand	Death Rate (1969) per Thousand	Rate of Annual Increase (%)
North America	19	9	1.4
Latin America	40	11	2.9
Africa	45	21	2.4
Asia	37	16	2.0
Europe	18	10	0.9
USSR	19	7	1.3

[1] United Nations Statistical Yearbook, 1970.

Data for most of the rest of the world is sparse, especially for the distant past. Death rates were higher and have been reduced more recently. Death rates of 7 to 11 per thousand are found in developed countries with higher rates for Asia and Africa in particular (Table 9.3). The greater excess of birth rate over death rate in those countries and in Latin America suggests that the birth rate diminution noted in Europe has not yet happened there.

In terms of methodology regarding a lowering of the birth rate, we find a large and varied number of techniques that have been devised. One group of techniques involves prevention of fertilization of the ovum with a spermatozoon. Abstinence from sexual intercourse, either on a permanent or periodic basis, the rhythm method, is in this category. So are mechanical barriers that prevent semen released from moving into the female reproductive tract. The most popular of these are the condom and the diaphragm. The technique of "coitus interruptus" or withdrawal of the penis before ejaculation is an old and rather ineffective method.

Another group of techniques involves using some method whereby the act of intercourse is not interfered with but in which the ovum or sperm are somehow prevented from normal functioning. The oral contraceptive or "the pill" is designed to prevent ovulation in this way. If the sperm ejaculated are inactive or infertile, then fertilization is impossible. Spermicides are used to kill normal sperm before they can reach the ovum. Another approach involves using douches that wash semen from the vagina. The ideal method here must be safe and reversible in the sense that fertility may be restored when use of the method is terminated.

Yet more drastic approaches are sterilization methods. For women the common approach is to tie off the Fallopian tubes leading from the ovary to the uterus. It is then impossible for sperm to reach ova when they have been released. For males the analogous procedure is the vasectomy, a surgical procedure in which the duct leading from the testis to the urethra is severed. The duct is known as the *vas deferens*, hence the name of the surgery. Both of these procedures are generally irreversible, and this is a severe disadvantage in the event that the individuals should find themselves in a circumstance in which they desire their fertility to be restored.

Another series of techniques involves normal fertilization followed by destruction of the resultant embryo. The use of an intrauterine device, which is a metal or plastic loop inserted in the uterus, prevents implantation of the embryo in the uterine wall. Abortion of an implanted embryo by surgical techniques is an old and controversial technique. More liberal attitudes and changes in the law in recent years in several states of the United States have further inflamed opinion both for and against the issue. The moral question of when abortion is acceptable, if ever, and when the life of the fetus comes under the protection of the law must be dealt with along with the practical fact that many women may be led to illegal abortionists and endanger their own lives. The issue is a most disturbing one for clergymen, politicians, physicians, and scientists alike. Moreover the question is going to become more difficult because the development of chemical means of abortion is imminent.

There is a final approach that should be noted even though it is not technically feasible. This is to institute contraception on a massive basis, by use of a contraceptive in the water supply, for example. The impact of the government in private sexual and reproductive affairs is something the future may bring, with its Orwellian overtones and implications if the population problem continues.

THE CHEMISTRY OF CONTRACEPTION

The female oral contraceptive has been the most important contribution of chemical research to the limitation of birth rate. Its effect has extended beyond its precise impact of the physiology of the female reproductive system. It is a social phenomenon. Primarily an influence in the developed countries beginning in the 1960's, some argue that it has changed marriage and premarital sexual behavior and had far-reaching effects on attitudes and institutions as well. In addition, some questions have been raised about side effects and long-term effects from the use of the pill. We begin our consideration of these concerns with a study of the normal functioning of the female reproductive system.

Sexuality exists in three levels or contexts. First there is the cellular level. It has long been known that differences in chromosomes direct differentiation of sexual characteristics. For the male, 1 of the 23 chromosomes in the gamete is truncated, smaller than the others. It is the Y chromosome and it is inherited from the father. For the female there is a corresponding chromosome of normal size, the X chromosome, also inherited from the father. The genetic characteristic of each cell nucleus, whether it has a specific sexual function or not, will bear this sexual imprint. Some genetic diseases, especially affecting males, are transmitted through the sex chromosomes because of the small size of the Y chromosome.

At the level of the whole organism, characteristics differentiating male and female operate on nearly every system of the body; in mammals they affect bones, muscles, hair distribution, pitch of voice, stature and body configuration, and behavior, to name a few. These distinctions are in addition to the obvious ones of the reproductive systems themselves. All these traits are known as secondary sex characteristics. At a third level, one sees sexual differences that are best described as psychological rather than physiological. These are imposed by the society as a whole and will differ from one culture to another. Thus dress will differ, or a certain courtship practice will be acceptable in one culture and unacceptable in another. Sexual practices and roles in a society will change with time as well.

The influence of chemistry on sexuality is primarily at the level of the secondary sexual characteristics. These are based upon the action of compounds secreted primarily by the gonads (ovaries or testes). Upon secretion they exert an effect on various organs and systems that convey typical male or female characteristics, and they also direct the operation of the reproductive system. These compounds are known as the *sex hormones*, and they have a structural similarity in both sexes. They belong to a class of organic compounds known as the *steroids*. All steroids share a common structure—four rings of carbon atoms fused together (Fig. 9.1).

The sex hormones differ from one another by the presence of different functional groups in the ring system at various positions. The principal male sex hormone is called *testosterone* (Fig. 9.2). Beginning at puberty it is secreted by the testes and circulated throughout the body. Its presence induces the maturing of the genitals, the growth of body hair, deepening of chest, increase in musculature, enhancement of the sex drive, etc.

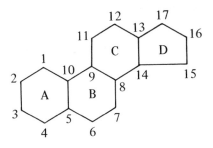

FIG. 9.1 The steroid ring system.

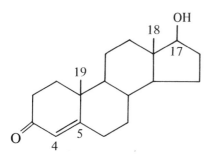

FIG. 9.2 Structure of testosterone.

In other species other secondary sex characteristics are manifested. The comb of the rooster is an example. One standard test for androgenic compounds is to inject the compound into a capon (a castrated rooster) and measure the enhancement of the comb that the compound causes.

Young adult men have concentrations of testosterone of about $5 \, \mu g/100$ ml of blood. Considerable amounts are excreted in the urine, to the extent of about 1 mg/day. This quantity is regenerated by further testicular secretion.

The female sex hormones are also steroids. There are two classes of hormones in females. The estrogens are responsible for female secondary sex

FIG. 9.3 The capon at the left showed extensive enlargement of the comb due to the injection of a male sex hormone for 22 days. (From Robert K. Callow and Alan F. Parkes, *Biochemical Journal*, vol. 29, page 1422, 1935.)

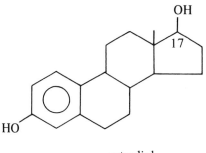

FIG. 9.4 Note the structural similarities to testosterone, the OH
group at C-17, and an oxygen
function at C-3 (Fig. 9.2). Note
also the differences. Ring A has
become aromatized or benzene-
like, and this has required that the
methyl group at the AB-ring junction be lost.

estradiol

characteristics. The hormone estradiol is the most important of the estrogens, and it is secreted by the ovary. One of the principal functions of the estrogens is to stimulate ovulation by concerted action with other hormones.

Another class of hormones is involved in the female reproductive system in mammals. These are the substances designed to foster the maintenance of the developing fetus in the uterus. They are the pregnancy fostering hormones or progestins. The most important of these is the substance progesterone.

The function of the progestins is to maintain an adequate blood supply to maintain the uterus in a healthy state to receive an implanted fertilized ovum and thus to maintain adequate nutrients, oxygen, and waste removal for the fetus through pregnancy. The principal organ secreting the progestins is a transitory one called the *corpus luteum* that forms as a "yellow body"—hence the name—at the ovarian site from which ovulation had taken place.

The estrogens and progestins are mutually antagonistic in effect. They tend to function in opposite ways but at the same time their actions are tied together and they are also linked to the pituitary gland where two other

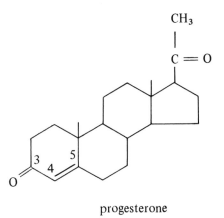

FIG. 9.5 Note again the similarity
of this structure to testosterone.
The only difference is the existence
of the two-carbon acetyl side chain
at C-17.

progesterone

hormones are produced that act as triggers for the ovarian hormones and in connection with them cause the menstrual cycle.

The pituitary hormones are proteins. One is follicle-stimulating hormone, or FSH. Its function is to stimulate the formation of a follicle at the ovary. The other is luteinizing hormone, LH. After the release of a mature ovum from the follicle, LH stimulates the formation of the corpus luteum at the point of follicle rupture. There is evidence that part of the brain, the hypothalamus, is also involved in the cycle, triggered by the gonad hormones and stimulating in turn the alternate production of these *gonadotrophic hormones* (LH and FSH) in the pituitary. It is also of interest that FSH produced in males triggers the secretion of testosterone in the testis as it does estradiol in the ovary.

The pituitary hormones all have similar structures. There are two subunits consisting of proteins and to which carbohydrate units are bonded. One protein unit, the α unit, may have the same structure in all gonadotrophic hormones, possibly even for different species. The β subunits apparently have different amino acid sequences for the different hormones and thus are the factor responsible for their different physiological activity. The precise structures are not yet known.

Figure 9.6 illustrates the complex relationships among the hormones in the female reproductive cycle involving the ovary, hypothalamus, and

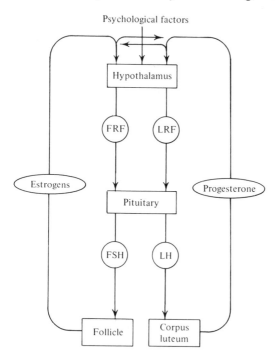

FIG. 9.6 Hormonal interactions in the menstrual cycle. (Taken from Clive Wood, *Human Fertility: Threat or Promise*, Funk & Wagnalls, New York, 1969, p. 67; published by arrangement with Thames & Hudson International Ltd., London.)

pituitary gland. In terms of the physiological effects, one can subdivide the menstrual period according to the hormonal actions that are occurring. It is the interaction of these hormones, estradiol, progesterone, FSH, and LH, with one another that causes the menstrual cycle, that causes ovulation, and that enables the reproductive system to foster and encourage conception and implantation of the fertilized ovum on the uterine wall.

The interaction of the hormones works this way. FSH causes the formation of an ovarian follicle. This follicle produces estradiol. Estradiol inhibits FSH production in the pituitary. As FSH declines, LH begins to be secreted. This causes ovulation and the formation of the corpus luteum. The corpus luteum produces progesterone, which inhibits LH. Progesterone also builds up the uterine blood supply. As LH declines, the corpus luteum atrophies. This cuts progesterone production. As this declines, the uterus finds itself with a surplus blood supply and this is sloughed off leading to menstrual flow. With all hormones then at low levels, FSH production is enhanced at the pituitary and the cycle begins again. Thus the cycle depends on the hormones turning one another off at the proper moment. There is evidence to suggest that two other hormones link the hypothalamus organ into the hormonal network. The production of the so-called "release factors" FRF and LRF are thought to trigger FSH and LH secretion in the pituitary.

In pregnancy it is necessary to maintain the uterine blood supply. Upon implantation of the fertilized ovum, the placenta produces a gonadotrophic hormone called *human chorionic gonadotrophin* (HCG). Its function is to substitute for LH and enable the maintenance of the corpus luteum beyond its normal time thereby continuing the production of progestins. Later the mature placenta assumes this function from the corpus luteum. With the continued production of progestins, FSH generation in the pituitary is suppressed, no ovulation occurs, and menstrual bleeding does not happen. Most important, the blood supplied to the uterus provides a rich nutrient source for the young embryo. Thus the missed period is the first sign of pregnancy. The urine of pregnant women is rich in HCG and its presence is the basis of most pregnancy tests. When the urine is injected into a female animal, HCG induces ovulation in that animal and this is positive evidence of pregnancy.

The Oral Contraceptive

The point of the modern oral contraceptive is to prevent ovulation from occurring. In order to do this, what amounts to an artificial period is created. The artificiality is created by the administration of artificial progestins and estrogens. The central event upon which subsequent ova depend is follicle stimulation. If no follicle forms, no ovulation occurs. Moreover, no corpus luteum forms since a follicle is a prerequisite to the corpus luteum. FSH suppression in the early phases of the cycle is required and estrogen is used to accomplish this since it suppresses FSH. Progesterone is also included in some formulations at this phase. The inclusion of progesterone in the pill formulation means that normal responses to the corpus luteal hormone is simulated. Thus the uterus develops a blood supply as if the corpus luteum were functioning. Perhaps it is accurate to portray the effect of the pill as a mimicker of pregnancy. Thus the increased progesterone and estrogen supply is seeming to tell the body that pregnancy has occurred. Defenders of the pill cite this as a positive feature of its use. By using a natural function (pregnancy) to advantage, the disruptive effects are eased, they contend.

Most oral contraceptive formulations differ. One type is known as the *combination pill*, which is a mixed formulation of steroid compounds, some of which are estrogenic and others are progestogenic. The combination pill is administered for 20 to 22 days followed by 6 to 8 days without pill usage. The other type is the *sequential pill* in which the two types of hormones are used in sequence. In the early days of the period, an estrogen-only formulation is used followed by a progesterone–estrogen mixture later in the period.

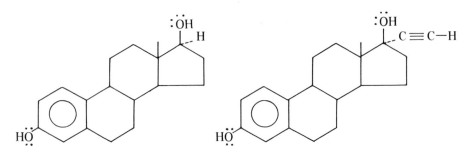

estradiol   17-α- ethynylestradiol

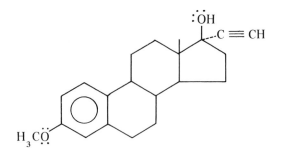

mestranol

FIGURE 9.7

Estrogens act to inhibit FSH early, while the protestogens later substitute for the action of the corpus luteum.

In the main, the compounds used are steroids, like natural estrogens and progestogens. For instance, compare the structures of estradiol with two common contraceptive estrogenic components. You will notice that 17-α-ethynylestradiol differs only by the presence of the carbon–carbon triple-bonded ethynyl group at carbon-17. The dotted line indicates that this group is projected below the plane of the steroid rings. The hydroxyl group at C-17 is above the ring. In mestranol the hydrogen of the OH group at C-3 is replaced with a methyl group.

A similar situation operates with the progestogenic contraceptive components. Again it is instructive to compare the most prevalent natural progestin (p. 230) with the contraceptives. Each of the contraceptives differs from progesterone but in different ways. Norethindrone differs in the substituents at C-17 and lacks the methyl group between rings A and B (note that the solid line present there in progesterone is missing in norethindrone). With medroxyprogesterone acetate, an acetate group replaces the hydrogen at C-17 (in progesterone) and there is a methyl group projected below ring B.

In the course of the developing technology of contraception, many new compounds have been discovered, synthesized, and put into use in different parts of the world. Almost all share with natural sex hormones the steroidal structure. A prominent exception is diethylstilbestrol, an estrogenic compound which is discussed for its role as a component in cattle feed and which may be a health hazard when present in meat (p. 245).

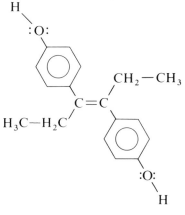

diethylstilbestrol

The next question is to discuss what is known about how the contraceptives work. There is undoubtedly more than one site of action. The most important, probably, is suppression of both gonadotrophic hormones from

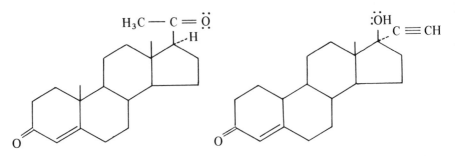

progesterone   norethindrone

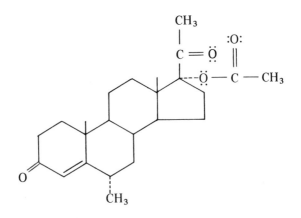

medroxyprogesterone acetate

FIGURE 9.8

the pituitary. From what we know of the natural menstrual cycle, this is consistent with the hormone-mimicking hypothesis. Since estrogen suppresses FSH in the critical first step toward ovulation, the cycle is subverted before it can get under way. The use of progestins later will suppress LH action. Perhaps it is unnecessary from the contraceptive point of view, but it does restore a kind of normalcy to the cycle by enriching the uterine wall so that normal bleeding occurs after the pill is stopped. Cessation of the pill is analogous to the atrophy of the corpus luteum.

There is some evidence that other physiological functions are also affected by some contraceptive compounds. One possible source of contraceptive action is upon the ability of the sperm to fertilize an ovum. A process called *capacitation* happens to sperm in the Fallopian tube before fertilization. Capacitation enables the sperm to synthesize an enzyme called hyaluronidase that is used to penetrate the egg cell membrane. Without capacitation there is no penetration. There is some evidence to suggest that some contraceptives inhibit capacitation. Another possible effect is on the receptivity of the endometrium to implantation of the incipient embryo, which depends on hormonal balance and which may be disrupted. The effects of hormones on cervical mucus is also being studied. During the ovulation phase the cervical mucus is very thin and sperm penetrate easily. During the rest of the cycle, this mucus is rather thick and not receptive to sperm. This is obviously a defense against infection that is maintained, except when sperm penetration is mandatory for fertilization. The cervical mucus is affected in some mammals by some contraceptives. It remains thick and tacky through the whole cycle.

The female sex hormones exert profound effects upon women, far beyond the reproductive system. To the extent that contraceptives mimic natural hormones they can also exert other physiological effects. Some of these are important, others are observable but not serious. Two areas for concern about the oral contraceptive have arisen: one is the possibility of cancer induction; the other is the effect upon blood clotting.

It has been shown that there is a small but statistically significant increase in blood clotting among women taking some contraceptives. This means that there are a few individuals who may be more susceptible to the clotting effect and who should not use the pill because of danger of internal blood clotting. Physicians can detect these individuals fairly easily and solve the difficulty by switching to another contraceptive method.

The question of whether some contraceptives may be carcinogenic is a difficult one. There has been no increase in the incidences of cancers of any kind among pill users compared with women who are not receiving the contraceptives. On the other hand, one must be concerned with the possibility of long-term effects. No one can say with certainty that administration of contraceptives for 20 years may not lead to higher instances, perhaps only slight, of some forms of cancer. It should be stressed that no evidence exists on either side of the question. No one has proved or disproved that possibility. There have been a few reports of difficulties under heavy dosages of contraceptives among animals. One product was withdrawn from the market in the United States in 1971 after a correlation of mammary cancer among beagle bitches given heavy dosages was established. That does not mean that normal dosage constitutes a hazard, but it is sufficient cause for concern that the law required its discontinuance.

Investigations in this and similar fields involving toxicity are very difficult. One cannot observe long-term effects in a short time. Small changes in statistical data are difficult to evaluate. To extrapolate from animals to humans is speculative, but one cannot justify human experiments until animal results are conclusive. The fact that the adverse effects occur under heavy dosages does not mean that these effects will be observed under normal lower dosage. Most physicians, pharmacologists, and chemists who have worked in the field take the position that the present oral contraceptives are safe. Most women who take the pill seem to prefer a small risk to an unwanted pregnancy. Many dangers of the pill, such as the clotting problem, are also present during a pregnancy, as it happens.

This is not to say that we should regard the oral contraceptive, combined or sequential, administered daily through most of the menstrual cycle for 20 years or so, as the last word in contraception. There are disadvantages involved, and there are other approaches that we can anticipate in the next few years. The next section discusses these aspects.

CONTRACEPTION IN THE FUTURE

Besides the risks, even though they may be small, there are some other disadvantages of the oral contraceptive. It is more expensive than simpler methods. It requires a degree of discipline and remembering that may be difficult for anyone who is not used to medication. Other systems of delivery of an antiovulation contraceptive and other types of contraception may well replace the pill as it is used presently in the United States and elsewhere.

Research is going forward in a number of different areas in connection with new approaches to contraception. Among these are

1. *The reversible vasectomy.* Even when a man wants no more children, he can conceive of circumstances that would change this. Divorce and remarriage or the death of wife or present children are such possibilities. Once the vas deferens is severed, its surgical reconstruction is required in order for sperm to be transferred from the testis. This surgical reconstruction is difficult at present and only about a 10% chance of success exists. Chance for improvement of this percentage is good however, with the improvement of surgical techniques.

2. *The development of improved intrauterine devices.* Research in this area centers on design of new materials and new shapes that will resist occasional problems of bleeding and expulsion. Among the interesting developments in materials for IUD's has been the fact that copper metal seems to have strong spermicidal properties. More research is also needed into the question of the physiological action of the IUD. Aside from the fact that its action is mainly centered on preventing implantation of the fertilized ovum, little is known about the specifics of this action.

3. *Development of a male oral contraceptive.* Developments in this area are much less advanced than in the female counterpart. In part this is due to less knowledge about normal male reproductive biochemistry and in part

because of a feeling that men are less sexually "responsible" than women or that they have more reservations about temporary loss of fertility. Such a drug, as with female contraceptives, would have to proceed through rigorous testing in animals before it could be used, and there would be similar doubts about side effects and long-term effects in the male pill as for the female pill. On the other hand a couple could gain some advantage by alternating the contraceptive roles. This would reduce the chance-taking for both by allowing normal reproductive function for each half the time.

4. *The development of long-term pills.* One of the big disadvantages of the oral contraceptive could be overcome if the daily regimen of pill-taking were not required. Thus the idea of a monthly or even less frequent administration has attracted interest. In this field, developments aimed at a time-release mechanism is the most likely approach, with efforts being directed toward a technique of implanting the medication under the skin with constant slow release following for an extended period.

5. *The after-intercourse pill.* A potential for a chemical abortofacient has emerged from research on a class of compounds known as the *prostaglandins*. These compounds were first discovered in the 1930's as the result of the observation that human uterine muscle tissue underwent a strong response to some substance in human semen, specifically in the plasma, or carrier fluid. Prostaglandins were first associated with the prostate, which is the basis for their name.

Because they are present in very small quantities, and because there are several different prostaglandin compounds in seminal plasma and other tissues, several difficult problems confronted chemists. The first was to isolate the prostaglandins from some tissue. While the richest source known is human semen, it is difficult to obtain large enough quantities to work with. The concentration is about 300 μg/ml, which is fairly high, but there are 13 related prostaglandins present, so that the concentration of any one is smaller still. The material selected for the initial chemical study was the seminal vesicles of sheep. During the 1950's the prostaglandins were isolated in pure form from this source.

While there are several different compounds, like steroids, they share several structural features. These are

1. A five-membered ring:

2. Two long carbon chains attached to adjacent positions on the five-membered ring, in most instances with *trans* (opposite) geometry as shown by the solid and dotted lines:

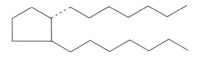

3. Oxygen functional groups at the next adjacent positions in the five-membered ring. For example, the prostaglandin PGE_1 has a ketone and alcohol group:

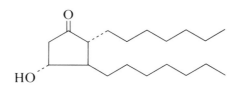

4. A carboxylic acid group at the end of one of the chains; and frequently a hydroxyl group and carbon–carbon double bond in the other.

Other prostaglandins are very similar structurally. For instance, one known as PGF_1 has the carbonyl group converted to an hydroxyl group. Otherwise the structure is the same as PGE_1. Others differ in similar ways, frequently by one or more additional double bonds in the chain or the rings. In order to determine structures it was necessary to extract prostaglandins from natural sources and to purify each from the others present in the extract. Then identification of the presence of each of the structural subunits including the problem of determining stereochemical orientations was necessary.

Since their initial discovery, prostaglandins have been found in numerous tissues. Human menstrual fluids are a rich source, but other organs remote from the reproductive system have been sources. Among these are liver, brain, eye, and kidney, which testifies to the wide occurrence of this class in nature. The most important physiological effect of the prostaglandins is upon smooth muscle tissues of all kinds. Interestingly, the effect may be either contraction or relaxation of the muscle. The physiological role in

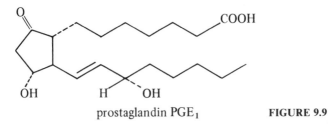

prostaglandin PGE_1 **FIGURE 9.9**

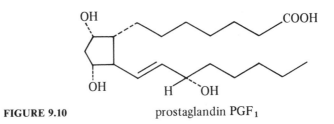

FIGURE 9.10   prostaglandin PGF₁

reproduction is probably the effect upon the uterus, which is, after all, a muscle. Some prostaglandins, in particular the PGE types, cause relaxation; others, the PGF types, seem to cause contraction and stiffening of the uterine wall. The presence in semen may relate to easing the passage of sperm beyond the uterus. PGE_1 is the active prostaglandin in semen and, by relaxing the uterine wall, contractions that would tend to expel the semen are prevented. Chances for a successful conception are thereby enhanced.

When exposed to PGE_1, the pregnant human uterus undergoes an increase in contractions just the opposite response from that of the non-pregnant uterus. The possibility exists that proper administration of a prostaglandin in adequate concentration could cause uterine contraction and expulsion of a newly implanted embryo. This is the basis for a once-a-month pill and the effect has indeed been observed. Thus, we have the potential for a chemically induced abortion.

Serious side effects arise because of the wide effects of prostaglandin on smooth muscle tissues. Muscles in blood vessels walls are expanded causing a drop in blood pressure suggesting applicability in treatment of hypertension. The digestive system contracts causing vomiting and/or diarrhea, and this is sufficiently unpleasant that oral administration for contraceptive purposes is probably ruled out. However, a vaginal suppository or tampon might be an effective method of administration for contraceptive purposes.

In order to produce the prostaglandins for widespread use as contraceptives or in other pharmaceutical roles, the synthesis of the compounds is required because not enough is available from natural sources. Synthesis of several prostaglandins has been accomplished within the last 5 years. Before their use as abortofacients or otherwise, problems of production and animal testing will have to be vigorously pursued. The possibility of a once-a-month chemical agent that can cause abortion of a newly implanted embryo appears imminent.

SUMMARY

The answer to the population crisis does not lie in one solution. Depending on the culture, the economic level, and other factors a different solution may be necessary. For some the vasectomy and for others a female contraceptive, mechanical device, or control of frequency of intercourse may work. New techniques and compounds are emerging that will expand the effectiveness of the total effort. Applied with good sense and compassion they can divert a population crisis in the opinion of most.

The social influences on reproductive physiology are little understood, except that some appear to be present. Under crowded conditions the sexual activity and reproductive rates of rats diminish. Some biologists believe that there may be changes in hormonal balances that cause decreases in sexual activities. Some even theorize that such a factor, operating independently of contraceptive use or abortions, may be at work in the world at present and will tend to work against at least an astronomical increase in population. Whether this is so or not, it is true that the birth rate of the United States has taken a sharp downturn in the 1960's and 1970's. From a post-World War II level of 26.6/1,000 population in 1947, the rate has declined to a low of 15.8 in 1972. This low value means an average family of 2.145 children, which is approximately that required for zero-population growth. Whether the reasons are increasing acceptance of abortion, a decreasing interest by families in having children because of the expenses involved, increasing independent feelings of women, more effective use of contraceptives, or more subtle social factors is not understood.

SUGGESTED READING

1. Gregory Pincus, "Control of Conception by Hormonal Steroids," *Science*, **153**, 493–500 (1966). This article describes the early development and testing of the first female oral contraceptives.

2. Egon Diczfalusy, "Mode of Action by Contraceptive Drugs," *Journal of Obstetrics and Gynecology*, **100**, 136–163 (1968).

3. Carl Djerassi, "Birth Control after 1984," *Science*, **169**, 941–951 (1970). A discussion of future contraceptive prospects.

4. Peter Ramwell and Jane E. Shaw, *Prostaglandins, Annals of the New York Academy of Sciences*, **180**, New York Academy of Sciences, 1971. An extended treatment of prostaglandins and their prospects.

5. Paul R. Ehrlich and Anne H. Ehrlich, *Population, Resources, Environment*, W. H. Freeman and Company, San Francisco, Calif., 1970. A discussion and analysis of all aspects of the developing overpopulation problem.

6. Clive Wood, *Human Fertility—Threat and Promise*, Funk & Wagnalls, New York, N.Y., 1969. A wide-ranging discussion of sexuality, contraception, and population problems.

7. Frederick S. Jaffe, "Toward the Reduction of Unwanted Pregnancy," *Science*, **174**, 119 (1971). An analysis of present use of all types of contraceptive methods and how one might go about improving the mixed results under present practices.

10

Extra Chemicals in Our Food

Ask why

Sign in the office of Kenneth W. Kirk
Former Associate Commissioner
Food and Drug Administration

A SURVEY OF THE FOOD ADDITIVES

We have already looked at the food-processing industry in terms of its contribution to biochemical oxygen demand through waste foods lost during milling, canning, cooking, cleaning, and other industry operations (p. 122). Now we turn to the question of some of the changes in food that manufacturers make, sometimes for economic or legal reasons and sometimes because they perceive the change to be in response to consumer demand.

Put yourself in the place of a food packer or processor for a moment. His problems are quite different from the steelmaker or petroleum refiner. There is the problem of raw material supply. Most crops are in season only briefly, but the demand for them is nearly year-round. Then one must contend with the problem of protecting the food from bacterial contamination and spoilage. The product must also be protected from oxygen, possibly for an extended time on the grocery shelf. Then there are problems of trying to meet consumer demand in terms of taste, convenience, appearance, and nutritional value. At the same time, government requires that the food products adhere to set standards for nutrition and purity. Finally, most of

237

these conditions change as consumer preferences, the laws, and agricultural patterns of production change. These factors lend complexity to management and decision-making in the food industry. The pattern in the twentieth century has been to rely increasingly upon chemical additives to change the properties of food in ways that help meet one or more of the demands.

There are many different kinds of additives used. We shall mention several types here. *Emulsifiers* are used to disperse one liquid in a finely suspended state into another. In foods they are used to disperse an oily or high-fat liquid into an aqueous medium. By keeping the oil suspended in water, the desired texture and homogeneity are maintained. Milk is a natural food emulsion. Emulsifiers are heavily used in dairy products and confections. Among emulsifiers the most used are derivatives of natural fats. Diglyceryl stearate is similar to triglyceryl stearate (p. 83), except that one hydroxyl group is not esterified.

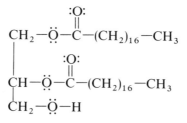

diglyceryl stearate

While the molecule is fat-soluble, the presence of the free hydroxyl function serves to increase the affinity of the molecule for the aqueous phase. Thus, the emulsifier can help hold fat particles of microscopic size suspended in water.

Some emulsifiers used occur in nature. Lecithin, found in egg yolk and in many other animal tissues, is sometimes used in the emulsifier role.

Acidulants, added to give a sour or tart flavor, are in significant use in soft drinks and gelatins. Citric acid, other organic acids, and phosphoric acid are frequently used.

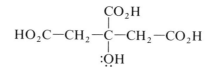

citric acid—a tricarboxylic acid

Stabilizers and thickeners are used to change the texture of a product. Ice cream is thickened with a derivative of cellulose (p. 21), carboxymethylcellulose, which has an affinity for water such that crystals of ice do not form.

As a result, the texture is smoother and the appearance of the ice cream is more pleasing.

Food flavorings are added to enhance the taste of foods. Salts and spices are in this category, but so are synthetic flavorings; for example, benzaldehyde is frequently used for cherry flavor.

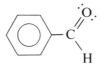

benzaldehyde

Many esters have fruity flavors; in fact, they are the natural constituents of many fruits. Both natural and artificial fruit flavors are used. Another important flavoring agent, monosodium glutamate, will be discussed at length later in this chapter.

Leavening agents make breads rise. Derivatives of phosphoric acid, such as calcium dihydrogen phosphate, $Ca(H_2PO_4)_2$, will react with sodium bicarbonate in a proton-transfer reaction.

$$H_2PO_4^- + HCO_3^- \longrightarrow HPO_4^{2-} + H_2O + CO_2 \text{ (gas)}$$

The CO_2 is trapped in the dough and makes the dough rise.

Preservatives prevent the growth of bacteria and consequent molding or spoilage. Salts of carboxylic acids, such as sodium benzoate, are frequently effective.

Antioxidants are used to prevent oxidation and hence rancidity in fat-containing foods such as cooking oil and potato chips. Two antioxidants, butylatedhydroxyanisole (BHA) and butylatedhydroxytoluene (BHT), are used frequently.

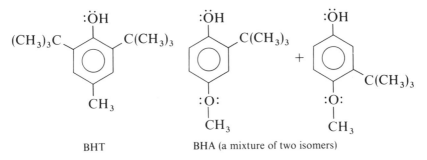

BHT   BHA (a mixture of two isomers)

The role of the antioxidants is to react with O_2, before it can react with fatty acids and other oxidizable compounds in the food. The antioxidants form free radicals by loss of the phenolic hydrogen. The radicals serve as scavengers for O_2, and by reacting with it they remove it from the system. This action slows the oxidative degradation of the foodstuff.

The *artificial sweeteners*, saccharin, and the cyclamates, will be discussed shortly in this chapter (p. 241).

Artificial colors are added to food to make the appearance more acceptable. Nearly 3 million lb of food coloring are added to foods per year. Some are natural; others are synthetic. Several dyes have been banned or restricted in the past by the government because of safety questions arising from results of testing with experimental animals.

Vitamins and other nutrients are added to improve nutritional quality. Normally vitamin enrichment is used in foods where that same vitamin occurs naturally. Thus, vitamin A is added to dairy products; the B vitamins, to grain products; vitamin C, to fruit juices, and vitamin D, to milk. Ironically, vitamins and many other nutrients are dissolved out of foods in cooking and other preparation operations during food packing.

Many critics of the food industry, among the health food or organic foods advocates, condemn the industry on this count and also on grounds that the additives are harmful. We must be careful to distinguish between the additives, however, remembering that some are natural materials. Citric acid, sodium benzoate, and the vitamins are examples. It is difficult to see how consuming these additives is harmful, since they are part of our ordinary diet and our metabolism has long since adapted to them. If the concentrations were far different from natural levels, a hazard is possible, but this is not the case. The synthetic additives such as BHA and BHT and the artificial sweeteners are another matter, however. In this case, mankind has not had much experience on the evolutionary time scale with the additives. There is thus more reason to expect problems. We shall look in more detail at four cases of controversial additives, but first we turn to the question of the law and the role of the Food and Drug Administration.

Farmers grow their own food, or at least they used to. What excess they produced, they took to the nearest town and sold either to the storekeeper or in a farmers' market to the consumers directly. This changed with the development of cities and industrialization, and an industry grew up concerned with transportation, processing, and distribution of food. Abuses developed, particularly in meat-packing. This led to an epochal muckraking work, *The Jungle* by Upton Sinclair, which depicted abuses. In the reform era of the Theodore Roosevelt administration, it led to the Pure Food Act of 1906, which was the first effort to control food shipped in interstate commerce. Its principal feature was periodic inspection of meat carcasses. In 1938, the Federal Food, Drug, and Cosmetic Act gave the Food and Drug Administration authority to forbid the use of additives. The trouble was that the FDA authority came after the fact. A company could use an additive in foods, and the burden of proof of lack of safety lay with the FDA. The FDA was required to base its action upon its own tests, which took time and more manpower than FDA was able to budget for that purpose. Moreover, court litigation posed further delays in the removal of the additive. In several instances FDA action caused the removal of dangerous additives from foods, but the procedure was too laborious in practice, and it did not cope well with the increasing numbers of additives that were being added to foods in the post-World War II era.

In 1958 an amendment to the 1938 act was enacted by Congress. Known as the Food Additives Amendment, it changed the burden of proof from the FDA to the company. In other words, a food additive could not be used until its safety had been proved. A similar act was passed in 1960 applying specifically to color additives. While companies must now submit test data to FDA, the agency itself still engages in substantial testing.

The 1958 act provides for three categories of additives. The first is the so-called GRAS list for additives. GRAS is an acronym for the first letters of the phrase "generally recognized as safe". The list includes common seasonings and spices and many additives; about 600 compounds are on the GRAS list. There is another category of substances that is approved, but not GRAS additives. In this case FDA sets limits on concentrations of the additives permitted. Finally, there are substances that are not permitted in any concentration. Of particular concern are substances that, under testing, are found to be carcinogenic in tests with animals.

The status of an additive may be changed by FDA action upon further testing. Some recent changes have been the cause of considerable public concern. We next examine some additives that have been under such recent scrutiny.

THE ARTIFICIAL SWEETENERS

For a long time it has been considered desirable for persons needing to limit caloric intake to restrict sugar consumption. For many years the acceptable alternate was *saccharin*,

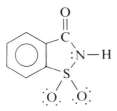

a compound that is at the same time an amide derivative of a carboxylic acid and a sulfonic acid. Saccharin is sweeter than sucrose by 300 times, but it has the disadvantage of leaving a slight bitter aftertaste.

In the late 1930's a new compound having a sweet taste but without the aftertaste was discovered by accident. Like saccharin, it is the amide of a sulfonic acid, called a *sulfamate*.

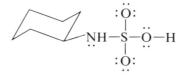

When a proton is lost, the compound is converted to its conjugate base.

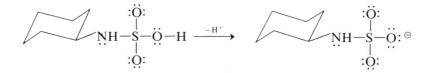

The conjugate base in the form of its calcium salt has been marketed under the name *cyclamate*. The discovery happened because of a flood. The disastrous Ohio River flood of 1937 damaged much of the Kentucky tobacco crop, which was curing in barns at the time of the flood. Later that year a graduate student in chemistry at the University of Illinois, Michael Sveda, switched from smoking his pipe to cigarettes because good pipe tobacco was not available. One day while in the lab, Sveda, who had a habit of chewing the cigarette in the manner of his pipe, ingested some of a compound with which he was working as he brushed tobacco shreds from his lips. The compound was a cyclamate, and its noticeable sweet taste led immediately to its exploitation as an artificial sweetener.

Although there were some early reservations concerning cyclamates, they were placed on the GRAS list and remained there until 1969. Late that year cyclamates were removed from the GRAS list and forbidden for use in foods because tests conducted by the manufacturer showed a high incidence of bladder cancer in rats injected with cyclamates at comparable concentrations to that used by some humans. At the same time, two other test results at FDA laboratories were also released. One showed appreciable deformities among baby chick embryos when the eggs were injected with cyclamates. This is an important study in that it means that cyclamate might exert a stronger adverse effect upon a developing fetus. Pregnant women, frequently advised to diet by physicians, consumed disproportionately more cyclamates than the overall population. Another study showed some chromosome damage in animals fed cyclamates.

A significant element in the cyclamate metabolism may be the fact that the cyclamate can hydrolyze, forming cyclohexylamine,

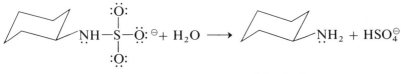

cyclohexylamine

This product is known to be a carcinogenic substance and this may cause the effects noted.

The removal of cyclamates from foods drastically changed the commercial soft drink business, which sells 250 million cases of low-calorie soft drinks per year. Most replaced cyclamates with saccharin again.

Similar doubts have been raised about saccharin, and if it should be restricted, the low-calorie sweetener user would find himself with no satisfactory alternative. Several alternative sweetners have been studied. One is a protein consisting of only two amino acid units, aspartic acid and alanine, that has been found to be much sweeter than sucrose. Also, a natural sweet protein has been isolated from a species of African wild berries. This compound is 3,000 times sweeter than sugar, and any adverse effects are doubtful, since the protein would degrade like any other in the intestine. The structure of the sweetener, called *monellin*, is not known. Until its structure is known, the synthesis will not be possible, and the only source will be isolation from the berries.

MSG—THE ALL-PURPOSE FLAVOR ADDITIVE

MSG stands for the first initials of *mono*sodium glutamate, which has the structure

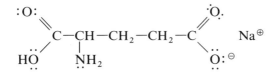

As you can see, it is the conjugate base of glutamic acid, which is one of the amino acids found in proteins. The "mono" prefix distinguishes the

compound from disodium glutamate in which both carboxylate groups are converted to their carboxylate salts.

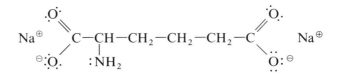

MSG is used in many prepared meats, vegetables, and seafood products because it imparts a pleasurable taste to meats and other protein foods. It is used in heavy amounts in Oriental foods, being first used in Japan. MSG differs from cyclamates in one significant respect as an additive, that it is a substance in abundance in nature. As such, it is not toxic. If heavy amounts were to be consumed, however, it is possible that some adverse reactions would appear.

One of the important food areas where MSG was used has been in baby food. The reason is strange. Infants do not have a sensitive taste, and so it is not important to improve flavors for them. Mothers sample the baby's food, however, and if they find it too bland, they will reject it. So to please the mother, baby foods contained MSG for some time.

Some experiments involving feeding MSG to infant rats have been reported in which the animals were injected with MSG at levels comparable to that fed infants. The results reported have conflicted. Some have pointed toward damage to brain tissues. Others have found no adverse effects as a result of MSG. While the value of MSG must now be questioned, especially where it serves no realistic nutritional or gustatory function, and where the concentration is high, there is no valid reason to ban its use in food in limited quantities. The use of MSG in baby food has been discontinued, however, because of potentially high concentrations and because of the uncertainty about whether a growing infant might be especially vulnerable. There are apparently some adults who have a special sensitivity and who must limit intake of MSG. The amounts should be clearly stated on labels so that these individuals are able to avoid excess intake. The decision of the FDA has been to retain MSG on the GRAS list, although it is possible that new findings may affect this decision.

DIETHYLSTILBESTROL

Diethylstilbestrol (DES) is something of an anomaly. On the one hand, it is an estrogenic compound, which means that it acts in the manner of a female sex hormone. Nearly all estrogenic hormones except DES are

steroids (p. 229). Its structure is

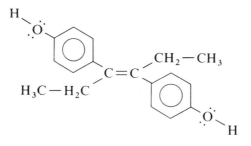

Its activity as a synthetic female sex hormone may be due to the phenolic hydroxyl groups, which are also characteristic of the steroidal estrogens.

In addition to its effects on the reproductive system, it was discovered that DES has the property of enhancing the rate of growth of livestock. Cattle fed small quantities of DES in their diet show faster growth rates when compared with animals fed standard diets. Since the animals are finished for market sooner (by 10% or so), it is economically attractive for cattle feeders, and as many as three-quarters of cattle slaughtered in the United States have been fed the compound.

Because DES has been found to be carcinogenic, its presence in food is forbidden by FDA. That means that cattle must be free of DES before slaughtering. Fortunately, DES is excreted fairly quickly, and it has been FDA policy to require that DES must be withheld from cattle for 2 days preceding slaughter. The problem has been enforcement. FDA has inspected only about 200 carcasses per year, and a few have shown traces of DES. While FDA has increased its carcass inspection to 6,000 animals, it seems likely that the inspection program is still lacking. The period of abstinence has been increased from 2 to 7 days prior to slaughter to try to reduce its effect. It should be noted that 21 foreign nations have banned use of DES in feed, and United States beef imports had been banned in some countries because of DES, until the United States also implemented a similar ban in 1972.

At the same time that the use of DES has been questioned as a *de facto* food additive, a prospective new use has emerged. DES has been found effective as a postcoital contraceptive. Pregnancy has been avoided when women ingested DES in small quantities following intercourse for which no other contraception method was used. At the same time there is evidence that pregnant women should avoid DES entirely. At one time in the 1950's, DES was used in large doses with a limited number of patients to prevent miscarriages during early pregnancies. In the early 1970's it was discovered that several daughters born of these pregnancies suffered from a rare form of vaginal cancer in their late teenage and early adulthood periods. The linkup between the cancer and DES usage has been established. Clearly, caution must be exercised in using DES in any context.

NITRITES

Sausages and other prepared meats have high fat contents, and as a result their color frequently appears as brown or gray instead of a ruddy red. Such color intensifies as the meat ages on the shelf. In order to maintain meat color, packers add sodium nitrite to the meat in processing. Nitrite ion reacts with myoglobin in meats to form nitrosomyoglobin. Myoglobin is similar to hemoglobin, and the reaction with nitrite involves the coordination with iron in similar fashion as that caused by O_2 (p. 174). Nitrosomyoglobin is responsible for the ruddy color. The nitrite ion also functions in an important role as a preservative preventing growth of *Clostridium botulinum,* the organisms causing botulism poisoning. In the acidity of the stomach, the nitrite ion gains a proton, converting to nitrous acid:

$$NO_2^- + H_3O^+ \longrightarrow HNO_2 + H_2O$$

nitrite   nitrous
ion   acid

Nitrous acid is reactive with secondary amines,* such as diethylamine, and can react to form a new class of compound called *nitrosamines.*

$$HNO_2 + CH_3-CH_2-\overset{\cdot\cdot}{N}-H$$

diethylamine   *N*-nitrosodiethylamine

The reaction has been shown to occur in small measure in experimental animals with various secondary amines occurring in the body. Nitrosamines have been established independently to be carcinogenic, so that at least a small chance exists that small quantities of nitrite, producing still smaller quantities of nitrosamines could lead to some cases of cancer which would otherwise not occur or which would not occur until later. Critics argue that since use of nitrite as a preservative requires lower nitrite concentration than the coloring function, the allowed quantity present in prepared meats should be reduced to only that needed in the preservative role. The FDA has refused to lower nitrite level maxima on the grounds that no nitrosamines have been observed to form in man from nitrites. However, nitrosamines have been found in nitrite-treated meats.

SUGGESTED READING

1. James S. Turner, *The Chemical Feast*, Grossman Publishers, New York, N.Y., 1970. This is the Ralph Nader Study Group Report on Food Protection and the Food and Drug Administration. It discusses the cyclamate and MSG decisions, and it is strongly critical of FDA and the food industry.

2. Howard J. Sanders, "Food Additives," Part 1, *Chemical and Engineering News*, **44**(42), 100–120 (1966); Part 2, *ibid.*, **44**(43), 108–128 (1966). These offer a good survey of food additives and the role of FDA in the food industry.

3. Howard J. Sanders, "Food Additive Makers Face Intensified Attack," *ibid.*, **49**(27), 16–23 (1971).
This reference discusses the more recent developments updating the discussion of Reference 2.

4. I. A. Wolff and A. E. Wasserman, "Nitrates, Nitrites and Nitrosamines," *Science*, **177**, 15–19 (1972). This surveys our current knowledge in this problem area.

II

The Chemical Crutches — The Drug Problem

Chemical Orientation In this chapter we discuss psychoactive drugs that are widely used in American society and in many other cultures for pleasure rather than for relief of pain. We begin by describing what is known of the chemistry of the nervous system, since this is the primary site of the action of the drugs. In preparation for this section the reader should review concepts of the nature of ions in solution and some aspects of electrochemistry. Also, the chemistry of amines should be reviewed. In connection with our discussion of amphetamines, the reader should consult references on the subject of optical activity and asymmetric carbon atoms. In the discussion of barbiturates a section devoted to an example organic synthesis is presented. Similarly a biosynthesis related to marijuana is included in that section. Sections of your text dealing with hydrogen bonding and with the activity of enzymes in biochemistry should be helpful in various contexts. If your textbook includes a section on alkaloids, it would be helpful reading in preparation for this chapter.

We deal now with the question of drugs; specifically those that have been of such controversy because of the fact that they are consumed not for reasons of health but for reasons of pleasure. We are talking here about the drugs of the new culture—marijuana and LSD; we are talking about the drugs of the establishment—alcohol and barbiturates; we are talking about the drugs of the desperate—heroin, morphine, and other addictive drugs. Our concern is with materials which alter our perception and our responses to the environment or which alter our basic functions such as sleep, hunger, and the sexual drive.

We are interested in these drugs because of their direct effects and because of possible side effects. Whether a drug causes harm in the short term is one question; whether it causes harm if used habitually for many years is another matter. The possibilities of causing genetic damage that could be inherited by future children, of inducing cancer, or of transmitting harmful effects to an infant in the womb must be of concern. In addition, one must worry about the possibility of addiction developing from use of a drug.

The objective will be to be objective, that is, to survey what is known and arrive at conclusions based upon these facts. This will not be easy because of the paucity of information. As a society, we have been derelict in studying these drugs until recently. Not enough funds have been allocated in this area by the United States or other governments and, consequently, too few scientists have been engaged in this research. In the light of so much public controversy in recent years, this problem is being eased now. The other problem in discussing this subject deals with the controversy of many of these drugs, and the passion with which many people hold views, both pro and con, on the subject. It is difficult to be objective in such an atmosphere, but commitment to the rational approach and the scientific method demands this of us.

THE NERVOUS SYSTEM

The effect of the drugs we are studying is upon the central nervous system. We believe in general that the drug mimics or substitutes for a naturally occurring agent and, in so doing, either inhibits a normal function or else causes the natural process to continue rather than stop when it should. Our first step then should be to explore the natural function of the nervous system, especially in terms of what is known at the molecular level.

The nervous system is both the governor of body functions and the connection between an organism and its environment through which the organism can most quickly respond to an external event. In man the development of the brain has led to a higher plane of nervous system activity not found in the other animals, at the level of rational thought and consciousness. The nervous system is complex anatomically, physiologically, and chemically. This brief sketch is designed to orient the reader to the later consideration of how some drugs interact with the nervous system on the molecular level.

There are two components, the *central nervous system* (brain and spinal cord) and the *peripheral nervous system*, consisting of nerves dispersed to all parts of the body. One part of the system is the autonomic nervous system, which functions independently of the will. Another deals with the area of consciousness and perception associated with thought process and with conscious response to stimuli from the environment. An ingested chemical substance is capable of affecting either or both systems.

The autonomic nervous system sees to it that all the organs of the body function in such a way as to benefit the whole organism. That is, it regulates respiration, increasing or decreasing the rate as the body's need for oxygen changes. Similarly, it will speed or slow the heart rate and increase or decrease the secretions of a gland as the needs for the gland's product changes. The functions within the autonomic nervous system are internally opposed. At the same time it must turn things on and turn things off. To accomplish this, two interconnected systems are used: one to stimulate the

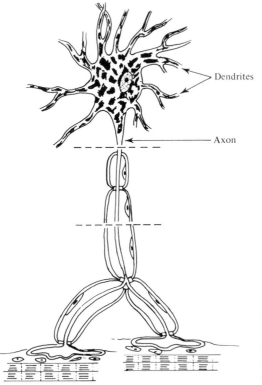

Dendrites

Axon

FIG. 11.1 A motor neuron. The impulse enters at the dendrite, travels across the cell body, then through the axon terminating at a synapse leading to another dendrite.

sympathetic nervous system and the other to inhibit the parasympathetic nervous system. By a balance of their opposing functions a dynamic equilibrium exists between the two.

Now let us take a closer look at the cellular level. Nerves are made up of cells called *neurons*. Structurally, neurons are shaped like long strings. A signal entering the neuron from the outside stimulates an impulse that travels to the terminus where it is sent on. The entering or receiving center is known as the *dendrite*. The terminus or transmitting center is the *axon* (Fig. 11.1). Each axon is connected to one or more dendrites so that one nerve can stimulate another. Thus, we have a network of neutrons constructed. A gap exists between the neurons, a very small gap called a *synapse*. In order for the functioning of the nervous system, two separate processes must occur. First, the impulse must be transmitted from one end of the neuron to the other. Then the impulse must cross the synapse from the axon of one neuron to the dendrite of another. Both of the processes occur by very interesting chemical reactions, which we are just beginning to understand.

The Sodium Pump

The transmission through a neuron is based on the fact that a small electric voltage or potential exists across the membrane separating the neuron from the extracellular fluid surrounding it. This means that there is a capacity for energy flow or work across the membrane. The existence of this potential is related to concentrations of two common cations in solution within the neuron and in the extracellular fluid. These are the sodium and potassium ions. The analysis of their concentrations in nerve fibers is of interest.

TABLE 11.1
CONCENTRATIONS OF IONS PRESENT
WITHIN AND OUTSIDE THE AXON OF THE
NEURON OF A SQUID[1,2]

Ion	Outside Concentration	Inside Concentration
	(mmoles/liter)	
K^+	10	400
Na^+	440	50
Cl^-	560	40

[1] Taken from O. C. J. Lippold and F. R. Winton, *Human Physiology*, Little, Brown and Company, Boston, Mass., 1968, p. 260.
[2] Similar values have been found for frogs and for mammals.

From Table 11.1 we see that within the nerve cell the concentration of K^+ is much higher than on the outside; the situation of the sodium ion is just the opposite. Note also that there is a distinct difference between chloride ion concentrations as well. In order to balance the positive charges, other anions, mostly derived from organic carboxylic acids, are present.

Differences in ion concentrations would not be what we would expect if the membrane on the neuron were not acting to disrupt the system. We would expect that the ions would flow across the membrane until both concentrations were equalized. The reason is that the flow of each ion across the concentration gradient would be faster in the direction from higher to lower concentration. The fact that this does not happen must be due to the fact that passage of the ions through the membrane is impeded in some manner.

When such a concentration gradient is in effect, we can measure a voltage if we place electrodes on opposite sides of the membrane, a rather delicate procedure. A voltage of about 70 mv (millivolts) has been measured across the neuron membrane. Potassium ion diffuses fairly well through the membrane itself. Its failure to do so when the neuron is resting is because the inside of the membrane has a negative electric charge and the outside of the membrane has a positive electric charge, so that the positive cation would have to do work against an electric field in diffusing to the outside. Na^+ should migrate into the neuron but does not do so because the membrane is relatively impervious to it.

The nerve cell at rest, then, is a small battery waiting to discharge its energy. The command that causes this discharge is a reversal of polarity due to a change in the permeability of the membrane to the Na^+ ion. Figure 11.2 traces developments in the course of nerve action. When the membrane becomes more permeable to Na^+, the ions flood across the membrane into the neuron. This causes the potential to drop across the membrane and is called *depolarization*. Next, the site of depolarization moves along the neuron. This happens by migration of ions in response to the fact that part of the inside is positive and part is negative. In this way the impulse travels from one end of the neuron to the other.

All this happens in less than a millisecond (10^{-3} sec). At the end, the potential has been deactivated and the nerve is incapable of further activity. The next order of business then must be the regeneration of the potential. To restore the 70 mv potential, the sodium ions that have entered the cell must be expelled. In the regeneration phase, energy is required; work must be done to create a new charge differential. This work is supplied by the ATP hydrolysis (p. 22) in similar ways to muscle fiber action. The permeability reverses, the sodium ion departs, and the resting mode is restored. Another impulse can then take place.

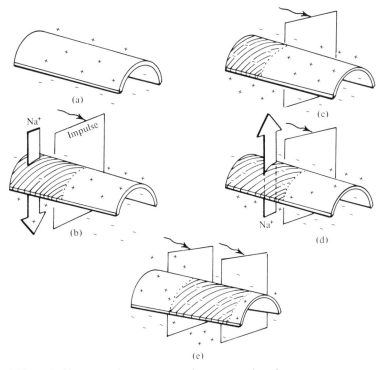

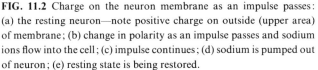

FIG. 11.2 Charge on the neuron membrane as an impulse passes:
(a) the resting neuron—note positive charge on outside (upper area)
of membrane; (b) change in polarity as an impulse passes and sodium
ions flow into the cell; (c) impulse continues; (d) sodium is pumped out
of neuron; (e) resting state is being restored.

The chemical structural change in the membrane, which is the key to
impulse generation and to regeneration of the resting potential, is not understood. The spark that sets this sequence in motion is also not known in its
detail, but the chemical agents responsible are known. They are formed in
the axon terminus of a neuron; they cross the synapse and initiate the impulse
phenomenon in the next dendrite. Thus, the impulse is shifted from one
neuron to another. The compounds responsible for this change are derivatives of ammonia known as *biogenic amines*.

The Synaptic Reaction

The nerve impulse is useless unless it leads somewhere. The connection
of one neuron to another is thus the central concern, and we shall deal with it
now. As the long axis of the neuron ends, it broadens into the synaptosome,

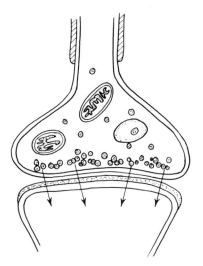

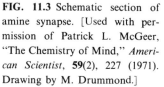

FIG. 11.3 Schematic section of amine synapse. [Used with permission of Patrick L. McGeer, "The Chemistry of Mind," *American Scientist*, **59**(2), 227 (1971). Drawing by M. Drummond.]

which is about 1 to 5 μ in diameter. The synaptosome contains numerous small spheres called *synaptic vesicles*. When an impulse reaches the synaptosome, the synaptic vesicles release chemical substances called *neurotransmitters*, which cross the synaptic cleft and activate the next neuron so that the impulse starts anew there (Fig. 11.3). Little is known about how the electrical impulse activates the release of the neurotransmitters nor how the neurotransmitters activate the impulse beyond the cleft, but the chemical substances have been identified. They are all organic compounds and all amines, and they are known collectively as the *biogenic amines*.

It appears that different biogenic amines act as the prime neurotransmitters in different parts of the nervous system. The sympathetic and parasympathetic systems in the autonomic nervous systems use different compounds. The central nervous system has still others. In the autonomic nervous system, the parasympathetic part uses an amine known as acetylcholine.

$$
\begin{array}{ccc}
CH_2 & \!\!\!\!\!\!\!\! & \!\!\!\!\!\! CH_2 \\
| & & | \\
:O: & CH_3-\overset{\oplus}{N}-CH_3 \\
| & & | \\
O{=}C & & CH_3 \\
| & & \\
CH_3 & &
\end{array}
$$

Note that acetylcholine is derived from the compound ethanolamine,

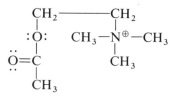

$$
\begin{array}{cc}
CH_2 & \!\!\!-CH_2 \\
| & | \\
:OH & :NH_2
\end{array}
$$

in which the alcohol group is converted to the acetate ester and the amino function is converted to the tetravalent nitrogen by three methyl substitutions. This develops a positive charge there just as in the ammonium ion, NH_4^+.

The principal agent in the sympathetic system is noradrenaline, also known as norepinephrine.

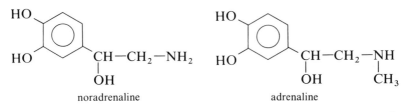

noradrenaline   adrenaline

As the name implies, noradrenaline is structurally very similar to adrenaline, but the latter has an additional methyl group on the nitrogen. Adrenaline is not produced in the nervous system but serves as an added activator from outside. Secreted by the adrenal cortex, adrenaline is produced under conditions of stress, and then it activates the sympathetic system.

In the brain two other amines function as neurotransmitters. They are *dopamine* and *serotonin*.

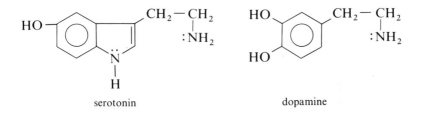

serotonin   dopamine

Note how many similarities these compounds have with one another and with noradrenaline. In each compound there is an aromatic ring with one or two phenolic hydroxyl groups. There is also a two-carbon side chain with a terminal NH_2 group. Dopamine and noradrenaline are both related to the phenolic compound catechol,

and are known, therefore, as *catecholamines*.

The molecular basis for the action of the neurotransmitter at the receiving neuron is not known and only speculation is possible. Since the compounds have so much structural similarity, it seems most probable that their action is similar. Changes in membrane permeability of the neuron are required, and there is probably a molecular site on the surface of the membrane that interacts with the amine. The likely molecular loci for this action are the OH groups which function as hydrogen bonders and the amino groups which can act as Lewis bases through the unshared electron pair (p. 171).

There is another important aspect in the picture. The neurotransmitter must be disengaged and deactivated; otherwise the signal will continue. It is as important to stop transmission as to start it. To accomplish this the amine is destroyed by chemical reaction. In the synaptosome one finds sites known as *mitochondria* that secrete an enzyme, monoamineoxidase (MAO). This enzyme converts the amine to the corresponding aldehyde. With dopamine, for example, the reaction is

The aldehyde product is inactive, and it is carried off into the extracellular fluid. Some of the amine is also removed by reabsorption by the synaptic vesicles.

With acetylcholine a somewhat different process is required to deactivate. The activity of the compound is terminated by hydrolysis of the acetate function by the enzyme acetylcholinesterase.

The reader should not be misled in thinking that the chemistry of the neurotransmission process is well understood. It is not. Much progress has been made, however, and undoubtedly will continue to be made in the area.

Drugs and the Synaptic Reaction

The psychoactive drugs we are concerned with are involved in the functioning nervous system. They alter perception, reasoning ability, motor coordination, and the body functions governed by the autonomic nervous system. Thus, we can assume that they somehow disrupt neuronal communications. Almost every drug we are concerned with is an amine, and some striking structural similarities with the biogenic amines exist as we shall see. The possibility that normal operation of the biogenic amine system is disrupted offers a hopeful approach to research in understanding the chemistry of the psychoactive drugs. One factor to keep in mind is that different drugs work in different ways. In the study of drugs it is of utmost importance to avoid overgeneralization. What is true for one compound may be false or unproved for another. To be sure, there are several ways in which an interference with normal transmitter metabolism can occur. As we consider each drug in this chapter, we shall look at what is known about its action. Among the possibilities for a drug acting on the synaptic transmission reaction are the following:

1. The drug exerts effects on the synthesis of biogenic amines in the synaptosome. The effect could be either to overstimulate, producing too much amine, or suppression, producing too little.
2. The drug affects the capacity of the synaptic vesicle to store the amines. For instance, a drug compound could compete for the storage sites on the synaptic vesicles.
3. Disruption of the communication of the impulse to synaptic vesicles could occur, so that neurotransmitters are not released or are not released at the proper time.
4. Competition for the active sites in the receiving neuron may be involved so that once the amine is released there is no place for it to go.
5. Inhibition of MAO or acetylcholine could prevent the removal of the neurotransmitter when that is necessary to cause the impulse to cease.

For each drug we shall consider the following questions:

(a) What is the active agent or agents?
(b) How does the substance exert physiological effects?
(c) Are there any long-range effects, such as addiction?
(d) What factors are not known about the drug?

From the chemical point of view, the molecular structures of the drug will be considered in assessing how they may function. In addition, one must take note of the metabolic fate of a drug in terms of chemical degradation. The physiological effect will be examined at the molecular level in the cell when possible. Further, there are such questions as the synthesis of the compound both in nature and in the laboratory, which will be of interest in some cases.

SUGGESTED READING

1. Patrick L. McGeer, "The Chemistry of Mind," *American Scientist*, **59**, 221–229 (1971). A review of what is known about biogenic amines.

2. Nikolaus Seiler, Lothar Demisch, and Herbert Schneider, "Biochemistry and Function of Biogenic Amines in the Central Nervous System," *Angewandte Chemie, International Edition*, **10**, No. 1, 51–66 (1971). This is a more advanced treatment of the same subject treated in Reference 1.

3. J. W. Phillis, *The Pharmacology of Synapses*, Pergamon Press, Inc., Elmsford, N.Y., 1970. This is a fairly advanced treatment of the synaptic phenomena.

THE ROLLERCOASTER: AMPHETAMINES AND BARBITURATES

Up—The Amphetamines

The amphetamines are amines that bear similarities in structure to adrenaline. Most commercial amphetamines are formulations of one or the other of two compounds.

The compound known as *amphetamine* or *benzedrine* has the structure

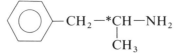

It is important to consider the structure at the carbon noted by *. An important feature is that there are four different substituents there. It is possible to orient the four different substituents for the tetrahedral carbon atom in two arrangements in space. For benzedrine these structures are shown in Fig. 11.4. This difference is due to the asymmetry at that carbon center. Note that the two structures are not superimposable. Does this mean that the structures represent the same or different substances? If we measure their physical properties, we find that melting points and boiling points, spectroscopic properties, and many aspects of chemical reactivity are identical. There are only two important differences. One is that the compounds have a property known as *optical activity* because of their asymmetry, which means that they interact in opposite ways with polarized light. This asymmetric property of the molecules makes them the same in the same way that one's right and left hands are the same—and different in the same way that one's right and left hands are different. While their shapes, bond distances and angles, and electron densities are the same, their spatial orientation make them mirror images of one another.

Let us continue the analogy of optical isomers and hands for a moment. Consider what happens when two people shake hands. While this is done normally with the right hands, if the left hands were used, there would be no

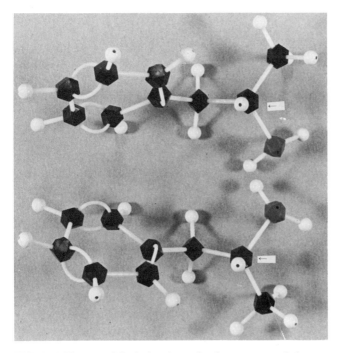

FIG. 11.4 These models depict the molecular structure of the two optical isomers of amphetamine. The structure at the bottom is the biologically active form *d*-amphetamine, or dexedrine. The top structure is the mirror image form *l*-amphetamine. The asymmetric carbons are shown by the arrows—the gray atoms are the amino nitrogens. Note that the two molecules are mirror images of one another, and that they are not superimposable upon one another.

difference in the action. The two left hands would link together in exactly the same way that the right hands do. Taken another way, the pair R–R and the pair L–L represent mirror-image pairs but are otherwise identical. If one person used his right hand and the other used his left, the situation would not be the same. To shake hands at all comfortably in this fashion requires one person to twist his wrist. The "fit" of the two hands is different. Again, the right–left combination could work either way, and we see two mirror images again, the R–L and L–R combinations.

Using this analogy, it is possible to understand the other significant difference in the properties of optical isomers—in their biological activity. Optical activity is a property that most key compounds of biology possess, including the carbohydrates, proteins, hormones, and nucleic acids. Two different optical isomers reacting with the same asymmetric molecule will do so differently. The reaction is like shaking another person's right hand with your right hand (R–R) and then with your left (R–L). The reaction is different in that same way. Thus, it is common for one optical isomer to exhibit a high biological activity while its mirror-image compound is much less active or even inert. The benzedrine isomers are examples of this. The isomer on the bottom, in Fig. 11.4, d-benzedrine, has much higher activity. One can infer from the fact that the site of interaction of the molecule in the nervous system is itself an asymmetric one, with which d-benzedrine interacts more effectively than l-benzedrine.

The usual commercial synthetic procedures which produce benzedrine give the mixture of d and l isomers (d and l refer to the directions of rotation of plane-polarized light). It is, however, possible to separate or *resolve** the mixture into the pure forms. The pure d-form is known as d-benzedrine or frequently as dexedrine. As stated earlier, it is the potent form of benzedrine because it is the active isomer in its pure form.

The second important class of amphetamines is very similar in structure to benzedrine. Pervitin is the mixture of the d and l isomeric forms. It differs from amphetamine in the presence of an additional methyl group on the amino nitrogen.

$$\langle\bigcirc\rangle - CH_2 - *CH - CH_3 \text{ (mixture of } d \text{ and } l \text{ forms)}$$
$$\qquad\qquad\qquad : NH - CH_3$$

pervitin

The d form is methedrine, and it also possesses the higher potency.

Note the structural similarity between these compounds and the fact that they are analogous in their differences to adrenaline and noradrenaline. In fact, methedrine is a contraction of the name methyl-dexedrine.

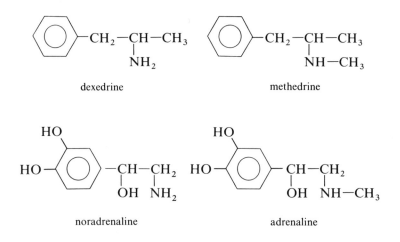

dexedrine

methedrine

noradrenaline

adrenaline

The relation of the amphetamines to adrenaline and noradrenaline clearly carries over into their effects on the nervous system, as they strongly affect the sympathetic nervous system in much the same way that adrenaline does. The most common effects at moderate dosage include increased respiration and blood pressure. Most users experience what they perceive to be a heightening of alertness, concentration, increased mental agility, and decreased fatigue. Sometimes the results are unpleasant, especially for heavier doses. Increased irritability, confusion, anxiety, and insomnia can result.

The structural similarity of amphetamines and the adrenalines and the physiological similarity of their effects could be due to one or more coincidental factors. Consider the possibilities: that amphetamines activate the same receptor sites that the natural amines do, that amphetamines stimulate noradrenaline production at the synaptic gap, that they prevent uptake of excess amine, or that the amphetamines disrupt the destruction process of biogenic amine at the synapse by competing for monoamine-oxidase. All of these would cause the overstimulation of the sympathetic nervous system that is observed. It has been established that the presence of an amphetamine will increase the amount of noradrenaline present in brain fluids, but the significance of this is not understood.

Beyond these immediate problems, amphetamines can cause two further difficulties. One is the phenomenon of the "speed freak," an individual who is using large quantities of the drug over a long period. Such individuals suffer severe physical and mental impairment. Withdrawal produces rather profound effects reversing the drug's impact; for example, voracious appetite and extended periods of sleep are common. While withdrawal effects are profound, they could not be described as addictive, as heroin and barbiturates are, for instance.

Amphetamines are frequently used in conjunction with other drugs, in particular with hallucinogens. Those who have tried the combination report that the amphetamine seems to enhance and prolong the induced hallucination. The use of two such drugs in combination is much more risky than the use of either singly, since drugs sometimes exert synergistic effects; that is, each contributes to the effect of the other such that the collective effect is more than the sum of the individual effects.

The use of amphetamines for various uses has received considerable national attention. A controversy has arisen from reports of their use by professional athletes to increase performance levels. The use of amphetamines for weight reduction by depression of the appetite is widespread; in fact, many doctors prescribe them. While there is a short-term success, the long-term effect is probably self-defeating because of other stimulative effects and because when the amphetamine is withdrawn, the appetite is enhanced again.

Perhaps nowhere is the idea of the relation between chemical structure and pharmacological activity better illustrated than in the barbiturates. The parent molecule for the compounds is barbituric acid.

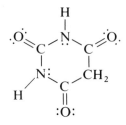

Note that the molecule consists of a six-membered ring containing two nitrogens. Three of the carbons are parts of carbonyl (C=O) groups; the fourth is as a methylene (CH₂) group. Barbituric acid is not a physiologically active compound. To induce barbiturate activity the hydrogens of the CH₂ group are replaced by alkyl (hydrocarbon) groups. The activity varies with the type of alkyl group used. The first barbiturate used was veronal, in which the methylene hydrogens are replaced by ethyl groups.

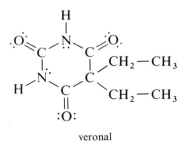

veronal

Among the more widely used barbiturates are:
    phenobarbital, where one group is the ethyl group; the other, the phenyl group:

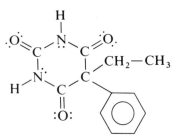

pentobarbital, also known as Nembutal, one substituent is ethyl; the other, a 2-pentyl group:

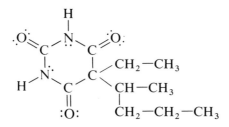

secobarbital, also known as Seconal, containing a 2-pentyl group along with an allyl group:

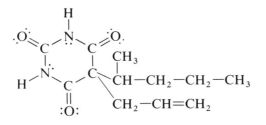

amobarbital, also known as Amytal, containing isopentyl and ethyl groups:

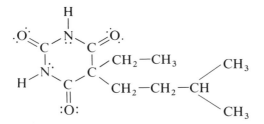

The effects of the barbiturates differ with differing alkyl substituents both with respect to the intensity of the barbiturate dosage and in the duration of the effect.

Barbituric acid, veronal, and the other barbiturates are not found in nature. They were originally synthesized in the laboratory and their pharmacological effects were later discovered. When veronal was found to be active, a systematic search was undertaken to prepare others and test them for activity. These compounds are a good illustration of the approach and the methodology of the synthetic organic chemist, and we shall take a moment to examine an organic synthesis.*

Some Chemical Comments

If we look at barbituric acid from the point of view of structure, we see that each nitrogen in the ring is flanked by two carbonyl groups. This structural unit is called an *imide*.

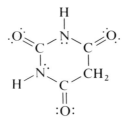

One may synthesize an imide by reacting an ester and amide. As a simple example, we use ethyl acetate and acetamide. Note that ethyl alcohol is the other reaction product.

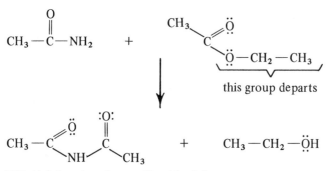

FIG. 11.5 Reaction of acetamide with ethyl acetate.

In order to synthesize barbituric acid we must cause the reaction to happen twice. Therefore, we must have a diamide and a diester as reactants. These are both common organic reagents that are readily available so that our choice seems a good one. The amide is urea; the ester is known as *diethyl malonate* or *malonic ester*.

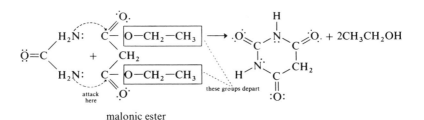

malonic ester

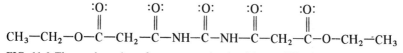

FIG. 11.6 The condensation of one urea molecule with two different malonic esters. Note that each malonic ester has a free end which could react with another urea, so that a polymer chain could be formed.

A final problem is to be sure that each urea condenses with the same malonic ester molecule twice. If each end of a urea molecule condensed with a different malonic ester molecule, we would have the beginnings of a chain not unlike that of a nylon polymer (p. 107). To avoid the problem is fairly easy. One simply runs the reaction in a dilute solution, so that the chances for a bimolecular reaction are reduced because the molecules are more separated by solvent.

To prepare veronal the methylene hydrogens must be replaced with ethyl groups. Starting with diethyl malonate there are two sequences of reaction possible. In one we substitute the ethyl groups onto diethyl malonate and then cyclize with urea. In the other the ethyl substitution occurs after cyclization.

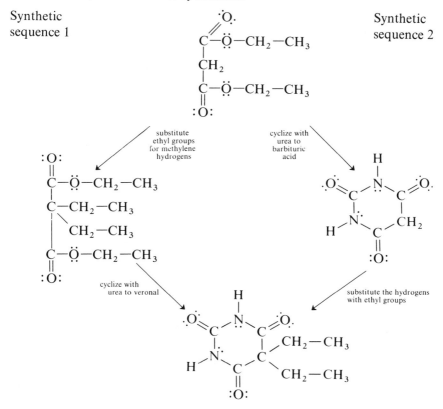

Synthetic sequence 1 is the better approach because it is quite easy to substitute ethyl groups onto diethyl malonate because of the fact that the hydrogens on the central carbon are moderately acidic, and a substitution reaction is accomplished in the presence of base. This synthetic approach and similar ones are used in the pharmaceutical preparation of the barbiturates.

The physiological manifestation of the barbiturates are in large measure the reverse of those that were encountered with amphetamines. They depress the activities of many systems of the body. The most important depression is exerted upon the central nervous system, such that they induce a deep sleep. In addition, the respiration rate is depressed noticeably. Under heavy dosage, heartbeat and other vital functions are depressed and this is the factor leading to death from barbiturate overdosage.

The exact physiological effect of barbiturates is not well understood, but at least two factors are thought to be involved. One effect is that barbiturates interact with neuroreceptors, especially those activated by acetylcholine. Thus, normal activation is impossible. The other possibility is that the barbiturates are interacting all the way along the nerve cell membrane such that the membrane permeability is not affected by normal activation at the synaptic junction.

Barbiturate usage leads to the development of tolerance if the user continues to use the drug. This leads to a serious problem because the difference between a lethal dose and an effective therapeutic dose is rather small for most barbiturates. As tolerance increases, the amount required to induce an effect increases, while the lethal dose level remains the same. Thus, this margin decreases, increasing the danger of a fatal overdose.

Barbiturates are addictive, meaning that habitual users suffer withdrawal symptoms of varying severity when they stop taking the drug. If the withdrawal effect is mild, there is simple irritability and insomnia. More severe withdrawal can produce convulsive seizures. Finally, there is the problem of the combined usage of barbiturates with alcohol. Both are depressants, and the overall result is greater than either taken singly. Thus, a person who follows a night of heavy alcohol consumption with a barbiturate is running a serious risk.

There are many legitimate situations for the use of barbiturates in treatment of numerous nervous system disorders and as a temporary sedative. In the United States we seem to have become overly dependent on the use of barbiturates when they are not needed on medical grounds. The use of both amphetamines and barbiturates is extremely widespread in the United States. About 60 million prescriptions for barbiturates are issued annually to 10 million users, which means that about 1 adult in 10 uses the drug at least on occasion. Amphetamines are used in comparable amounts for both weight control and as a stimulant. In addition to the amphetamines taken under a doctor's prescription, an equivalent quantity produced by pharmaceutical companies reaches the illicit market. Added to that is an unknown quantity that is synthesized in clandestine laboratories.

There are some serious questions here. Of course, we need to know more about the physiology of amphetamines and barbiturates, which means more understanding of the functioning of the compounds at the synapse at the molecular level. Then one must ask, has the medical profession been well-advised to encourage the use of these drugs given our lack of understanding of their function and the hazards that we know exist for their users? A reassessment of this policy seems to be in order.

SUGGESTED READING

1. L. S. Goodman and A. Gilman, *The Pharmacological Basis of Therapeutics,* The Macmillan Company, New York, N.Y., 1965. This discusses general aspects of both amphetamines and barbiturates.
2. *The Non-Medical Use of Drugs:* Interim Report of the Canadian Government's Commission of Inquiry, Penguin Books, Baltimore, Md., 1970.
3. G. M. Dyson and P. May, *May's Chemistry of Synthetic Drugs,* Longman's, Green & Co., Ltd., London, 1959. Discusses synthesis of principal amphetamines and barbiturates.

THE WORLD OF UNREALITY

Alkaloids

Among the more baffling and interesting activities of the plant kingdom is the production of compounds capable of exerting profound effects on the central nervous system of higher animals. The phenomenon may be a defense mechanism against predators, and while not many plants have developed this synthetic capability, the effects on the victims can be quite striking. Both the intensity and the nature of the effect vary.

Among compounds in this category are nicotine from the tobacco plant:

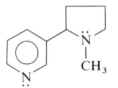

caffeine:

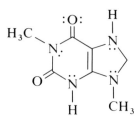

strychnine:

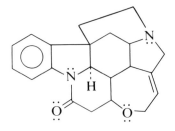

coniine (from hemlock):

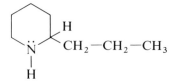

and cocaine:

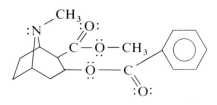

These compounds and a number of others share two structural character-istics: that they have at least one amine function present and that the amino nitrogen is present with carbon atoms in a ring. In some cases you will note that several rings are involved. Since the amino group makes the compound basic, these active plant derivatives are known as *alkaloids*.

The presence of the nitrogen is thought to exert an important influence upon the activity of the nervous system. In some cases the activity is un-doubtedly due to the fact that the compound either simulates or disrupts the activity of biogenic amines. A significant class of the alkaloids are the hallucinogens of natural origin. Indeed all the known major hallucinogens are alkaloids except the active agent in marijuana and the compound myristicine from the nutmeg plant.

The Hallucinogens

Among the important nitrogen-containing compounds are mescaline, obtained from the peyote cactus in southwestern United States:

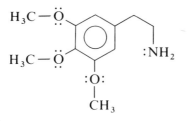

psilocybin, found in some mushroom species of Mexico:

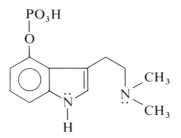

harmaline, from *Peganum harmala,* an Indian spice:

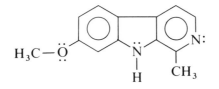

scopolamine—also known as "truth serum"—from the common jimsonweed:

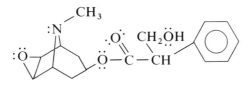

With the exception of scopolamine these naturally occurring plant hallucinogens bear noteworthy structural similarities to the biogenic amines. Note the correlation of the structures of psilocybin with serotonin, for instance.

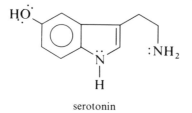

serotonin

In the same manner, mescaline is structurally similar to the biogenic amine noradrenaline (see p. 255). The most accepted interpretation of the bizarre effects of these compounds is that they mimic activity of the biogenic amines. Thus, they are frequently known as *psychotomimetic* drugs. Here again, much study is needed to understand further the physiological chemistry involved.

While these compounds are interesting for many reasons, it remained for a synthetic compound produced by an organic chemist—LSD—to provide the strongest and most controversial hallucinogenic effect. While this compound is synthetic, it is related to natural products. Lysergic acid is a complex polycyclic molecule.

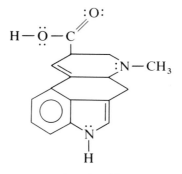

lysergic acid

Note that it also has the ring system of serotonin in the lower part of the molecule. The *acid* in the name refers to the carboxyl group in the top left part of the structure. The compound is not particularly acidic because the two amino groups confer basic properties to the molecule so that both weak acidic and basic groups are present.

Lysergic acid is produced by ergot, a fungus that grows in rye. It is not a psychotomimetic drug, even though its structure is not dissimilar to some plant hallucinogens. When lysergic acid is converted to the amide function, a compound of mildly hallucinogenic properties is obtained. Indeed lysergic acid amide and some similar compounds are found in morning glory seeds, and this was responsible for a "run" on the morning glory seed stocks of seed companies in the 1960's when this became known.

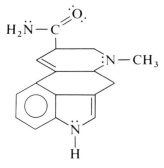

lysergic acid amide

The strongly hallucinogenic properties of LSD are conferred when ethyl groups are substituted for the hydrogens in the amide (Fig. 11.7). This is lysergic acid diethylamide. The German word for acid is *säure*. The initials LSD are taken from the German name lysergsäurediäthylamid.

LSD was first prepared by a Swiss chemist Arnold Hofmann in 1938 as a routine derivative of the acid. The compound was not of any particular interest, but 5 years later Hofmann accidentally ingested some LSD, something he would never have done deliberately. He shortly began to experience strange changes in perception and vision, in other words, the first LSD trip. That the compound was responsible for the hallucination was soon established, when Hofmann experienced another hallucination later under controlled conditions. The intensity of the experience was all the more striking because of the fact that so little compound was needed. The usual dosage to produce hallucinogenic effects is 50 to 100 μg (10^{-6} g).

The similarity to serotonin in structure is thought also to apply in the LSD effect. Why the structure around the carboxyl group plays such an important role is not clear. One possibility is that the diethylamide group causes the compound to pass readily through the blood-brain barrier in a way that the parent acid cannot.

Several adverse effects can be observed among LSD experiences. The most important and best documented is the "bad trip." While the usual effect of LSD is a positive experience, there is a significant minority of experiences in which the person has emotional trauma of fright and terror or he is led to dangerous irrational acts, even suicide. In effect, any person

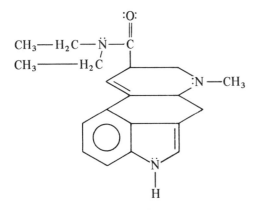

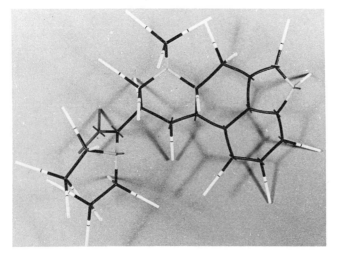

FIG. 11.7 Structure and molecular model for LSD. The diethylamide portion is shown at the left. The light gray centers are the nitrogens; the dark gray is oxygen.

ingesting LSD, even an experienced user, faces the possibility of a very unpleasant experience.

Several other charges have been leveled at LSD. Among them:

1. LSD causes genetic damage as manifested in breakage of chromosomes in cells.
2. LSD can cause serious birth defects upon an unborn fetus and damage to the reproductive organs and their capacity to produce healthy sperm or egg cells.
3. LSD can induce cancer.

These assertions are extremely difficult to prove. Cancer or other problems suffered by an LSD user may be due to some other cause entirely. The LSD, having been obtained by illegal means, may have contained an impurity because of improper or incomplete purification procedures. The user may have used another drug in addition.

Studies on animals have tended not to offer conclusive evidence in support of any of the assertions above. On tests carried out to date, comparisons of the results on test animals with untreated controls show little statistical support for these effects. On the other hand, this does not mean that these effects may not be observed for some individuals or that long-term effects may not develop. At this time there is clear risk involved in taking LSD until these questions have been answered. It is likely that research underway will serve to answer these questions with more certainty in the next 5 years.

SUGGESTED READING

1. D. H. Efron, *Psychotomimetic Drugs*, Raven Press, New York, N.Y., 1970. This book discusses all aspects of the psychotomimetic drugs, including chemical, pharmacological, and some clinical studies.

2. John Cashman, *The LSD Story*, Fawcett Publications Inc., Greenwich, Conn., 1966. This volume develops the historical and early clinical events of the drug.

3. N. I. Dishotsky *et al.*, "LSD and Genetic Damage," *Science*, **172**, 431 (1971). An analysis of evidence of side effects resulting from LSD.

4. David Solomon, ed., *LSD: The Consciousness-Expanding Drug*, G. P. Putnam's Sons, New York, N.Y., 1964. This book presents a large number of views on the LSD question as it stood in 1964.

5. Norman R. Farnsworth, "Hallucinogenic Plants," *Science*, **162**, 1086 (1968). A discussion of chemical and botanical roles of natural hallucinogens.

AMERICA'S MOST SERIOUS DRUG PROBLEM

Alcohol and man have had a long and vicissitudinal coexistence. The discovery of alcohol and its seeming ability both to exhilarate and relax is lost in antiquity, but it is easy to envisage how primitive man would stumble across the chemical reaction of the fermentation of sugars to ethyl alcohol by microorganisms. In any event, the use of alcoholic drinks of one kind or another is found in nearly every culture on earth. In the United States the pattern of use of alcohol has changed in the last century. A greater percentage of the adult population are alcohol users now, but the per capita consumption has dropped. About 70 million adult Americans use alcohol in varying degrees of moderation, and this is about three-fifths of the adult population. At the retail level, the sale of liquors constitutes a 7-billion dollar business.

If it is true that alcohol serves to relieve the stress that modern life demands, it is also true that there is a decidedly somber side of the picture. Alcohol seriously impairs the action of the nervous system, limiting to some extent judgment, perception, and motor function. This leads to serious impairment in the operation of an auto or any other task requiring a fair degree of coordination. The drunken driver causes an alarmingly high proportion of auto accidents because of this.

For some 5 million drinkers, the use of alcohol has led to a loss in the ability to control drinking habits. These are the alcoholics, and their fate is frequently to lose their ability to contribute in a useful way to society, followed by ill health and premature death. This addictive aspect of alcohol, as well as the normal metabolism of alcohol, is not well understood. We shall attempt to deal with these questions at the molecular level. Before that we shall briefly treat the question of the chemistry of ethyl alcohol.

Some Chemical Comments

The molecule that causes all this "ferment" is a small, simple one:

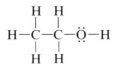

The hydroxyl group places it in the organic functional group category of the alcohols. The two-carbon chain means it is a derivative of ethane. It has commonly two names, ethyl alcohol or ethanol, and, of course, it is frequently called *alcohol*, although not usually in a chemical context, because of the ambiguity in that term.

Because of the presence of the hydroxyl group, ethanol is capable of hydrogen-bonding* interactions between ethanol molecules, with water molecules, or with similar molecules (Fig. 11.8). Because of H-bonding

(a)   (b)

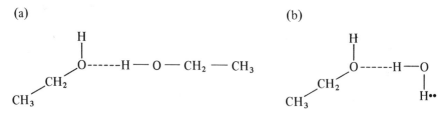

FIG. 11.8 Hydrogen bonding (a) between two ethanol molecules and (b) between ethanol and water molecules. In each case hydrogen bonding may involve more than two molecules in a more complex linkage.

interactions and their resultant stability, ethanol and water are soluble in one another in all proportions. Unlike water, there is only one H bond possible per ethanol molecule, and the intermolecular network is weaker and less extensive, and less energy is required to break down the network. This means lower melting points and boiling points for ethanol ($-114°C$ and $78.3°C$, respectively) than for water ($0°C$ and $100°C$, respectively). This is true in spite of the fact that the molecular weight of ethanol is considerably higher. On the other hand, a comparison of ethanol with propane, a hydrocarbon of about the same molecular weight with a melting point of $-187°C$ and a boiling point of $-42°C$, shows how significant the role of hydrogen bonding is for ethanol, since its melting and boiling temperatures are much higher. Yet, ethanol is still an organic molecule owing to its hydrocarbon part, and this factor makes it a good solvent for organic compounds.

The liquor industry is involved with the preparation of alcoholic beverages from plant matter. The first step in the process is *fermentation*, in which the sugars from the plant are converted to ethanol. In the second the fermented product is subjected to *distillation*, in which the ethanol is concentrated. Depending on the particular type and quality desired, a process of aging in which chemical changes are occurring slowly serve to enhance the liquor quality.

Chemistry of Fermentation

The starting material in the genesis of ethanol is natural sugar found in grains or fruits. In the case of those liquors that are derived from grain, the principal sugar is maltose, consisting of two glucose units, linked together (see p. 21), derived structurally from starch.

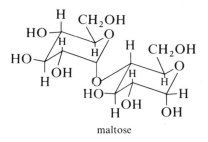

maltose

In the action of fermentation, the sugar is reacted with a plant microorganism in yeast, known as *Saccharomyces cerevisiae*. This organism produces an enzyme, maltase, which breaks maltose down to glucose.

$$\text{Maltose} + H_2O \xrightarrow{\text{maltase}} 2 \text{HO}$$

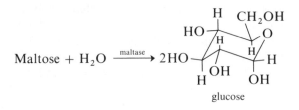

glucose

Then another enzyme present, zymase, plays a key role in effecting the conversion of glucose to two moles of ethanol and two moles of CO_2 through a complex metabolic pathway that may be summarized in the reaction

$$\text{Glucose } (C_6H_{12}O_6) \xrightarrow{\text{zymase}} 2CH_3CH_2OH + 2CO_2$$

The end product of fermentation is used directly in the case of beers and wines, which are typically 3 to 12% ethanol by volume. Whiskeys and brandies are, respectively, products of fermented grains and fruits in which the ethanol content is enhanced by distillation. Most of the alcohol distills as a mixture of 95% ethanol by volume with water. A few volatile flavor constituents from the fermentation also distill out and impart a characteristic flavor and color to the liquor. After distillation, the liquor is diluted to about 40% ethanol by volume.

Alcohol Metabolism

The intoxicating effects of ethanol begin as alcohol ingested passes into the bloodstream. At low levels, 0.05% blood ethanol by volume, one finds typically euphoria, loss of anxiety, dulling of the sense of humor and inhibitions, and sometimes an impulsiveness of action that, combined with coordination impairment, may be either amusing or deadly depending on the situation. At a level of 0.15% ethanol, the individual will probably become drowsy and he will experience vomiting. The level of alcohol sufficient to cause death is not much higher, so that vomiting is necessary, in effect, to shut off the ethanol supply to the blood, preventing further increase. This fact prevents ethanol from being an extremely toxic agent to man. At only 0.30% ethanol, impairment of respiration is serious enough to result in death.

The ultimate metabolic fate of ethanol is that the body extracts energy from the molecule in much the same way as it would from the parent carbohydrates. The net equation then results in the conversion to CO_2 and H_2O:

$$2C_2H_5OH + 6O_2 \longrightarrow 4CO_2 + 6H_2O$$

In fact, ethanol is a rich energy source; 1 mole of ethanol releases 328 kcal on combustion.

The first step in its metabolism is the oxidation of ethanol to acetaldehyde. This reaction takes place in the liver under catalytic action of an enzyme, alcoholdehydrogenase.

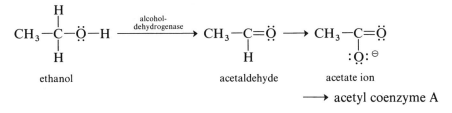

ethanol   acetaldehyde   acetate ion

$\longrightarrow$ acetyl coenzyme A

Acetaldehyde is subsequently oxidized as well, forming acetate ion that metabolizes by forming acetyl coenzyme A, a normal participant in the metabolism of carbohydrates. Since acetyl coenzyme A is involved in the series of reactions known as the *Krebs cycle*, whereby normal sugars are degraded to CO_2, providing energy for cellular metabolism; the ultimate breakdown of ethanol then proceeds along this normal route.

Both the oxidation steps above, leading from ethanol to acetate, require a second reactant, a coenzyme. As the alcohol or aldehyde is oxidized, the coenzyme is reduced, thus bringing the oxidation and reduction processes into balance. The coenzyme is nicotine adenine dinucleotide, or NAD. Thus, we represent what happens as two half-reaction processes. In the oxidation half reaction, ethanol is oxidized to acetaldehyde. In the reduction half reaction, NAD is converted to its reduced form, usually depicted as NADH. A reversal of the coenzyme reaction, occurring independently, restores NAD, which may then be used again.

One of the serious effects of chronic drinking is that the NADH/NAD ratio in the cells gets seriously out of balance in that too much NADH is present. This factor may lead to the degenerative changes in liver cells leading to liver cirrhosis, the most serious disease of alcoholism.

The effect of ethanol on the central nervous system may well involve the biogenic amines. Certain changes in the normal metabolism of serotonin and noradrenaline have been demonstrated. For example, the normal metabolism of serotonin is as follows:

1. Oxidation with monoamineoxidase (MAO) to the aldehyde:

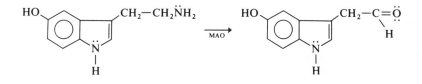

2. Oxidation of the aldehyde to the carboxylic acid 5-hydroxyindoleacetic acid (5-HIAA):

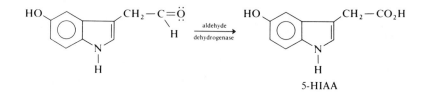

5-HIAA

The enzyme for this reaction is aldehyde dehydrogenase, the same enzyme as in the acetaldehyde oxidation. In this reaction also, the NAD $\longrightarrow$ NADH reduction occurs on the coenzyme as the reduction half reaction. If NADH is too high, the aldehyde appears to follow a reduction route, to the corresponding alcohol, 5-hydroxy tryptophol (5-HT-OH), in a reaction catalyzed by alcohol dehydrogenase,

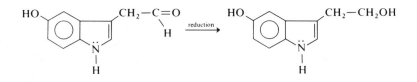

with an oxidative half reaction, NADH $\longrightarrow$ NAD occurring in the coenzyme system. By completely changing the metabolic route of decomposition of serotonin, it is reasonable to expect changes in serotonin concentrations in and around synaptic regions in the central nervous system and consequent aberrant functionings.

The NADH/NAD ratio factor may play a role in the habituating use that alcohol induces in many people. We shall now discuss the chemical basis of alcoholism.

Alcoholism—The Molecular Mystery

There is no consensus among experts on the question of the nature of alcohol addiction. The problem is—is it physiologically addictive or is it only a tragic habit whose dependence is psychological? If it is physiologically based, then we have some hope that its chemistry can be unraveled. Some evidence has been developed that suggests that alcohol addiction is physiological, hence chemical, just as morphine is. There is scientific evidence suggesting that aldehyde intermediates may be involved here as well.

In test-tube experiments designed to simulate *in vivo** conditions, some interesting reactions occur: in the case of the neurotransmitter dopamine, for example, a reaction occurs between the amine and its aldehyde metabolite, in which they condense together by eliminating a molecule of water.

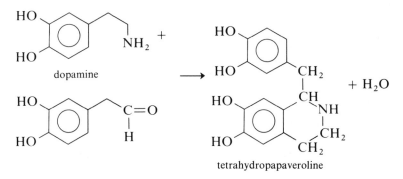

tetrahydropapaveroline

Note that the aldehyde oxygen is lost in the formation of the water molecule along with hydrogens from the nitrogen and the aromatic ring. The condensation product is tetrahydropapaveroline. It is known to be involved in the biosynthesis of morphine. There is a remote suggestion then that in the situation where the aldehyde persists near the synapses without further reacting in the normal oxidative way, that this reaction occurring between it and dopamine could lead to an addicting product. Before one can prove the chemical link between alcohol and morphine addiction, which this experimental result suggests, a great deal of research, especially with experimental animals, will have to be conducted. A similar condensation between the acetaldehyde produced from alcohol and a biogenic amine can also be observed; for instance, with noradrenaline:

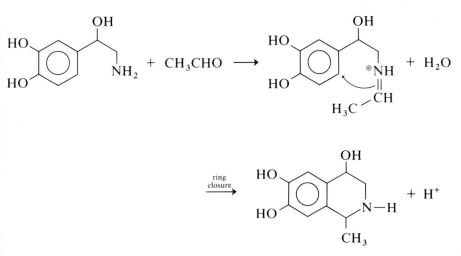

This product belongs in a family of alkaloids known as *tetrahydroiso-quinolines*. Members of this family have been found to be psychoactive as hallucinogens and in other ways. Again, much research and, in particular, *in vivo* studies are required before a definite chemical correlation can be established.

At present, this chemical evidence is not strong, but it suggests that an understanding of a physiological basis for alcoholism is within our reach. If we can understand that problem, we may be able to devise an effective treatment or, by understanding what makes some persons liable to alcoholism while most can avoid it, we can forewarn the potential victim and thereby avoid the tremendous price that both the alcoholic and society pay for his habit.

SUGGESTED READING

1. V. M. Sardesai, ed., *Biochemical and Clinical Aspects of Alcohol Metabolism*, Charles C. Thomas, Publisher, Springfield, Ill., 1969. This is a series of papers describing the known aspects of the biochemistry of alcoholism.

2. Berton Roueche, *The Neutral Spirit, A Portrait of Alcohol*, Little Brown and Company, Boston, Mass., 1960. A brief description of historical aspects of the interrelation of alcohol and man.

3. R. M. Glasscote, *et al.*, *The Treatment of Alcoholism, A Study of Programs and Problems*, Joint Information Service of the American Psychiatric Association and the National Association for Mental Health, Washington, D.C., 1967.

4. B. Kissin and H. Begleiter, ed., *The Biology of Alcoholism*, **Vol. 1**, *Biochemistry*, Plenum Press, New York, N.Y., 1971. A rather extensive discussion of all aspects of alcoholic metabolism, metabolism, and the preparation of alcoholic liquors.

MARIJUANA

Almost everyone has an opinion about "pot." Some consider it to be almost like alcohol, in other words, an intoxicant, whose use by minors and in connection with operating an automobile should be restricted, but little else. Others think marijuana is as dangerous as heroin, an addictive evil that will lead the user first to a life of crime to maintain his "habit" and later to an early grave. Still others see marijuana as part of the phenomenon of Eastern mysticism: that its use enables one to gain new perspectives and insights into his own existence, thereby deepening and enriching his life experience.

The euphoric and mildly hallucinogenic effects deriving from the species *Cannabis sativa* or hemp plant have been long known in many parts of the world. Individual plants possess only the male or female reproductive system, not both, a rarity in plants. During the period of maturing of seeds in flowers of the female, a resin is secreted that acts to protect the developing seeds by repulsing predators until full maturity of the seeds is achieved. This resin is known as *hashish* in the Middle East and in the Western world. If one grinds up the whole flowering top, seeds, leaves and all, the resin is naturally less concentrated. This latter form is what is normally termed *marijuana.* The method of ingestion is through the lung for either hashish or marijuana. The smoker inhales deeply, holding the smoke in the lungs as long as possible to maximize absorption of active compounds into the bloodstream. Ingestion through the digestive system is also possible, but the effects then develop more slowly because absorption from the intestine is a slower process.

While the psychoactivity of marijuana has been known for a long time, understanding of the molecular basis of its activity has been uncovered only in very recent work. The problem for the chemist has several facets, each developing from the previous one. We can summarize these in the following questions:

1. Is there a single compound or class of compounds responsible for the psychoactive effects of marijuana?
2. How can one isolate that compound or compounds from the plant and establish that it is the active agent?
3. What is the molecular structure of the compound?
4. It is usually necessary to synthesize the compound in order to prove the structure without question. How can one do this?
5. How does the plant synthesize the compound? In other words, what is the biogenesis of the compound?

We shall look at the marijuana question in terms of these chemical questions.

Isolation of The Active Agent in Hashish

When hashish is subjected to distillation at reduced pressure, it is found that the distillate that is separated from the nonvolatile parts of the resin possesses the pharmacological activity. Extraction of hashish with an organic solvent such as hexane or methanol removes the active compounds from sugars and other inactive resin constituents that are not soluble in such solvents. A complex mixture is obtained that can be separated into the pure compounds by chromatographic techniques such as elution from a column containing a solid of high surface activity. On such a column, compounds are retained in varying degrees depending on their polarity. By running a solvent through the column, the less polar ones are removed from the column quickly, the more polar ones later. By changing the receiving flask at the right time, the compounds may be individually isolated.

Several compounds of similar structure are isolated from hashish by these procedures. They are collectively known as *cannabinols*, taking their name from the species and from the fact that all of them have a phenolic hydroxyl group.

One of the first compounds whose structure was established was cannabidiol. Its molecular formula was found to be $C_{21}H_{30}O_2$. Catalytic hydrogenation adds 2 moles (four atoms) of H_2 to the molecule, producing tetrahydrocannabidiol, $C_{21}H_{34}O_2$. This shows that there are two carbon–carbon double bonds in cannabidiol.

The arrangement of the carbon skeleton is established by the fact that one can cleave the molecule into two parts, an aromatic hydrocarbon, *p*-cymene, and a phenol, olevitol, which is found in many plants.

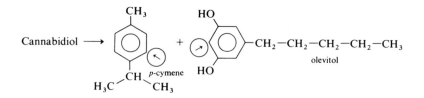

Since these two compounds have 21 carbon atoms, we have accounted for all the carbon structure of the cannabinol system. The position of attachment of each ring to the other is shown by arrows such that the carbon skeleton is

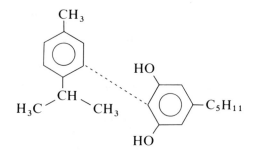

The ring on the left is not aromatic in cannabidiol. One of the two double bonds to which the two moles of hydrogen are added is in the three-carbon chain at the bottom of the structure. The other is in the ring.

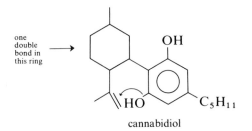

one
double
bond in
this ring →

cannabidiol

This establishes the carbon skeleton of cannabidiol except for the position of the double bond as shown on the left.

Upon treatment with dilute acid, a ring forms through one of the phenolic OH's with the three-carbon side chain (note the arrow in the cannabidiol structure above). The double bond is lost in this process. One double bond remains somewhere in the upper six-membered ring.

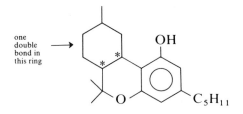

one
double
bond in
this ring →

The formula, $C_{21}H_{30}O_2$, is also that of the most active agent found in hashish. The determination of the position of the double bond in this substance was a difficult problem, and its solution required the use of a spectroscopic technique called *nuclear magnetic resonance* (NMR) spectroscopy, which can detect various structural features depending on the magnetic environment that the nucleus displays in a magnetic field. This compound is known as Δ^1-tetrahydrocannabinol or Δ^1-THC. (The double bond is in the position shown.) There is one further question of structure, the stereochemistry of the ring junctions, denoted by asterisks in the structure. Again using the NMR data, it is possible to show that connection is *trans*, that is, that both the hydrogens and the connecting carbons are opposite one another instead of on the same side of the ring. The "same-side" or *cis* geometry would give the arrangement

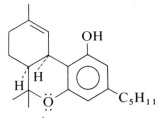

cis Δ^1-tetrahydrocannabinol

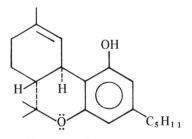

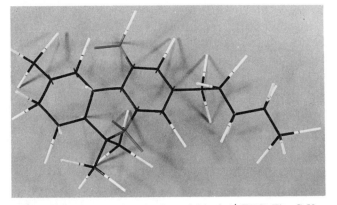

FIG. 11.9 Structure and molecular model of Δ^1-THC. The C_5H_{11} side-chain appears on the right side of the photo. The gray centers are the oxygen centers.

If we compare Δ^1-THC with cannabinol,

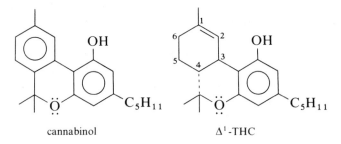

cannabinol Δ^1-THC

we can rationalize the name of the compound. The prefix *tetrahydro* indicates the addition of four hydrogens to cannabinol into the benzenoid ring of cannabinol on the upper left. The Δ prefix denotes that the one double bond is still present and the superscript identifies the numbered ring position where the double bond is found. The name Δ^9-THC is sometimes used both in scientific work and in the popular press for the compound. It originates from a different sequence in numbering the carbon atoms in the molecule.

Synthesis

In order to establish without question the structure of Δ^1-THC, a synthesis of the compound from known materials was carried out. Several successful approaches have given unequivocal confirmation of the proposed structure. The most straightforward synthesis of Δ^1-THC is a condensation reaction between olevitol and $(-)$verbenol, a naturally occurring terpene in plants. Three events occur in sequence. First, under the influence of an acid catalyst, the two reactants condense:

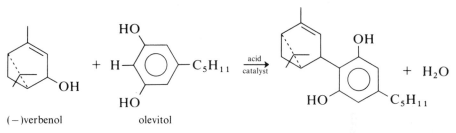

$(-)$verbenol   olevitol

Note that a molecule of water is lost as the verbenol becomes attached to the aromatic ring of olevitol. Now the third ring is formed by a rearrangement process involving the double bond, the verbenol bridge, and one of the OH groups on olevitol.

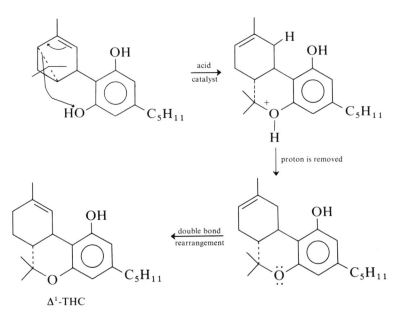

Δ^1-THC

The only change then required to produce Δ^1-THC is rearrangement of the double bond.

The synthetic compounds show biological activity similar to that of natural Δ^1-THC, further confirming that it is the natural compound.

Biogenesis*

In *Cannabis sativa*, the course of the biological synthesis is different from the chemical synthesis just described. Olevitol or the analogous carboxylic acid is undoubtedly involved, with the terpene geraniol a strong possibility for the other reactant [Reaction (1), Fig. 11.10]. Once the adduct is formed,

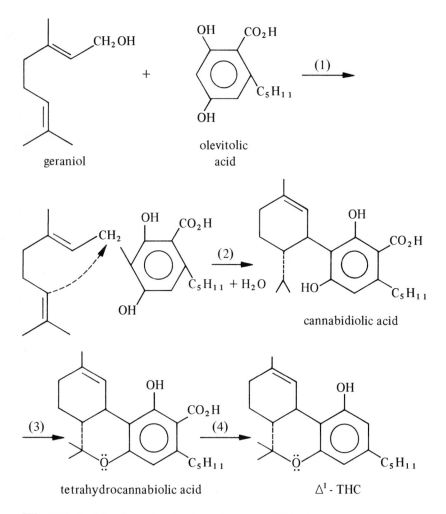

FIG. 11.10 Partial pathway showing formation of Δ^1-THC in *Cannabis sativa* plant.

a series of rearrangements close the rings [reactions (2) and (3)] to tetrahydrocannabinolic acid. Then loss of the carboxyl group produces Δ^1-THC.

The other compounds present in hashish resin arise either as precursers such as tetrahydrocannabinolic acid (the CO_2H group has not been lost) or as $\Delta^{1,6}$-THC, the rearranged double bond compound. From the presence of these compounds we can envisage a partial metabolic pathway for the formation of Δ^1-THC shown in Fig. 11.10.

It is known that hashish and marijuana from plants in tropical and subtropical regions are more potent than that from temperate regions. This is why smokers prefer the Mexican to the Kansas product, for example. The Kansas variety produces more cannabidiolic acid and less THC than the tropical type. The other factor in quality is the degree of "ripeness." Cannabidiolic acid is present in larger quantities in unripe plants, which points to its being a precursor in the biosynthesis of Δ^1-THC.

Physiological and Psychological Effects

There is a key missing link in our understanding of how THC acts upon the nervous system and other organs to exert its profound effects. That it does so is obvious, although proponents and opponents differ on the effects. At any rate, let us examine what is in fact known about this aspect of THC.

In short-term effects, an increase in the heartbeat is usually noted. Changes in the membranes in eye, nasal, and respiratory regions are observed and so is increased frequency of urination. Oral ingestion can cause nausea and diarrhea. Under very heavy dosage more severe effects involving coordination impairment are noted. The most common response to overdose is sleep, however, and no deaths from cannabis overdose have ever been proved or documented.

What about effects on the chronic user? In particular, is marijuana addictive? Does it induce cancer? Can it cause birth defects when taken by a pregnant woman? Or can it cause genetic damage? Here again we have the same problem as in the case of LSD. It is hard to prove or to disprove any of these allegations on a rigorous basis because they all involve sampling users who have had widely diverse experiences and backgrounds and who may have used other drugs as well. The other approach is to test animals, and there one has the difficulty of extrapolating the results to humans.

One aspect of nearly unanimous agreement among users and investigators is that the addiction, if any, is of a mild sort compared with heroin or even with alcohol. Most describe the dependence as psychological rather than physiological, which only means that those who smoke marijuana tend to like it and wish to continue to smoke it. On the psychological level, definition and measurement are difficult. The drug strongly affects mood, leading usually to tranquil, reflective attitudes. The user usually feels a heightened sense of understanding and perception. Although the data on the subject is rather limited, there is evidence supporting impairment of some intellectual and coordination performance. Most users would agree with this assessment also. Other reported characteristics such as heightened appetite, increased sexual desires, and a change in time perception have little hard evidence in their support but are commonly reported. No evidence exists that marijuana is physiologically addictive in the way that the opiates are. For example, no withdrawal discomforts are found when a regular user stops.

In the current controversy on the marijuana question, a certain perspective is important. For instance, events in the late 1960's and early 1970's repeat the experience of the 1930's, where marijuana usage in urban ghettos was very heavy. Perhaps the most serious criticism one can make is that while the paucity of information was realized then, no effort was made to establish the facts needed to deal with marijuana rationally. From all indications, the scientific community is now beginning to give marijuana the attention it deserves, and the U.S. Government and others are prepared to provide funding to get important answers.

Here are some of the key questions:

1. What are the products of THC metabolism? One compound, 7-hydroxy-THC, is known.

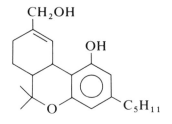

7-hydroxy-THC

The methyl group at the top left has been converted to a CH_2OH group. This compound is itself physiologically active. Some experimental results show this compound to be more potent than Δ^1-THC itself and very possibly the pharmacologically active agent.

2. What is the lifetime of THC and its metabolites? Some evidence on this question suggests a very long half-life in the body. This term refers to the period required for the drug to diminish in concentration by one-half. The physiological effects noticed do not persist, however. This persistence may be involved in the reverse tolerance phenomenon claimed by some users who find intensified effects with less material after experience with the drug.

3. What is the molecular basis for the pharmacological effects of marijuana? In particular, is its action related in any way to biogenic amine metabolism?
4. Are there legitimate medical uses for marijuana or for synthetic THC; for instance, as an analgesic, an antihypertensive agent, an antidepressant?
5. What synergistic effects occur when marijuana is taken concurrently with other drugs, such as alcohol or amphetamines? In other words, do two drugs taken together exert effects different from either acting singly?
6. What evidence is there for harmful effects caused by long-term usage?

While it is extremely important to know the answers to these key questions unequivocally, evidence is growing in support of the position that marijuana cannot be classified with either the hard hallucinogens or with heroin and the opiates. From 15 to 20 million Americans, more than 1 in 10 in the adult population, have smoked marijuana, and among college students more than half have tried it. While many of these users have been casual or only one-time smokers, it is clear that many Americans have chosen to violate the drug use laws. Of more importance is the position taken by the National Commission on Marijuana and Drug Abuse in 1972. After thorough study, the commission recommended an end to criminal penalties for private marijuana use or possession, although it favored misdemeanor penalties for public use or possession of more than one ounce, public marijuana intoxication, or operating an auto under the influence. Retention of felony penalties for cultivation or sale of the drug was recommended by the commission.

Whether relaxation to the extent recommended will occur in the near future is doubtful, given public positions taken by national political leaders. The next years should see the research results that will clarify the question of any dangers or side affects involved. One may also hope that a just resolution of the law based on this research may be possible in a cooler emotional climate on the subject.

SUGGESTED READING

1. R. Mechoulam and Y. Gaoni, "Recent Advances in the Chemistry of Hashish," *Fortschritte der Chemische Organische Naturstoffe*, (*Advances in the Chemistry of Organic Natural Products*), **25**, 175–213 (1967) (in English). This work treats the details of the chemical questions of hashish.
2. Andrew T. Weill, N. E. Zinberg, and Judith V. Nelson, "Clinical and Psychological Effects of Marihuana in Man," *Science*, **162**, 1234–1242 (1968). Discusses procedures for testing marijuana among human subjects.
3. Leo E. Hollister, "Marihuana in Man: Three Years Later," *Science*, **172**, 21–29 (1971). A review of current knowledge on the physiological effects of marijuana.

4. *The Non-medical Use of Drugs*, Interim Report of the Canadian Government Commission of Inquiry, Penguin Books, Inc., Baltimore, Md., 1970. This work discusses extensively drug usage in Canada and elsewhere.

5. Mayor's Committee on Marijuana : *The Marihuana Problem in the City of New York : Sociological, Medical, Psychological and Pharmacological Studies*, Jacques Cattell Press, Lancaster, Pa., 1944. Commonly known as the LaGuardia report, it is perhaps the best description of what is known about marijuana published, at least as to what was known at that time. Its conclusions were neglected until very recently.

WHITE DEATH: THE OPIATES

Modern United States has experienced a substantial increase in crime in recent years. There are surely numerous causative factors in this development, but most police and others concerned agree that high among these factors is the activity of heroin addicts, many of whom sustain their addiction by theft of property or by prostitution. The cost to the nation in loss of property is estimated at 8 billion dollars/year. The number of persons addicted to heroin in the United States is estimated at between 250,000 and 500,000. The problem is most serious in New York City and in other areas of high urban concentration. In recent years the problem has been found increasingly in the suburbs and even in rural areas.

In spite of the upsurge in heroin usage, the drug and its relatives are far from new. Derived from the poppy, their euphoric and analgesic effects were known in antiquity. The opium poppy produces opium from the juice of the unripened seed pod. No other part of the plant produces psychoactive substances. The opium is a dark, crystalline material prepared by evaporation of the oil. For many centuries opium has been in use in many parts of the Old World, especially in east and south Asia. The material could be either ingested through the lungs by smoking or sniffing or it could be swallowed and absorbed from the digestive tract. The British colonial rulers in the nineteenth century tended to encourage the use of opium for economic reasons, and it became so serious that it precipitated violence in the form of the "Opium Wars." The first use of opium in the United States was probably introduced by the Chinese. Opium was even used in patent medicines in the nineteenth century.

While addiction arising from opium is relatively innocuous, early work led to the discovery of the active components. Opium is a mixture of a number of alkaloids. The most active compound is morphine, whose structure proved to be exceedingly complex to ferret out and even more difficult to synthesize. The principal factor of difficulty is that the molecule has such a complex ring system (Fig. 11.11). The alkaloid character is due to the heterocyclic amino nitrogen.

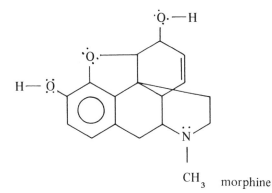

CH$_3$ morphine

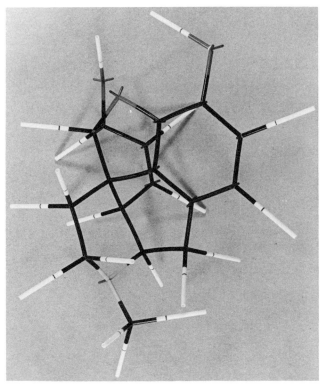

FIG. 11.11 Structure and molecular model for morphine. In the model the dark gray centers are the oxygen; the light gray is nitrogen.

In addition to morphine in opium there are about 20 other alkaloids whose structures have been identified. One of these is codeine in which one of the free OH groups on morphine is converted to a methoxyl group.

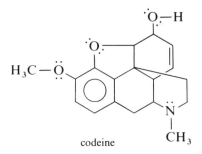

codeine

Codeine is a weaker narcotic than morphine.

Another class of alkaloids in opiums, known as the *papaverine alkaloids*, has a different structure. The papaverine alkaloids are less active narcotics than the morphine group.

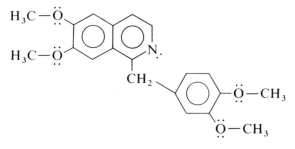

papaverine

The compound heroin is a synthetic derivative of morphine produced in a chemical reaction known as *acetylation* in which the acetyl group

is substituted for the hydroxyl hydrogens of morphine. This reaction can be accomplished by reacting morphine with the compound acetyl chloride.

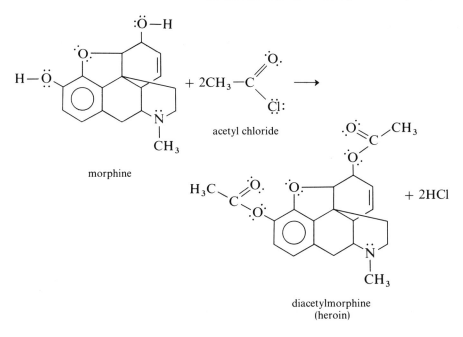

morphine

acetyl chloride

diacetylmorphine
(heroin)

The other product in the reaction is HCl. The narcotic effect of heroin is far stronger than that of morphine. The major and pronounced effect of the morphine narcotics is upon the central nervous system. The effect is to depress CNS activity as manifested in these results. First there is analgesia, the relief of pain, which is the principal legitimate use of morphine. Morphine is a very effective analgesic at dosage levels of 5 to 10 mg. Morphine and heroin also produce a feeling of euphoria, which is the principal attraction for nonmedical use. In addition, the subject becomes drowsy and his mental acuity and physical activity diminish. A second area of effect of morphine and heroin is the bowel. Heroin addicts are habitually constipated.

The basis of the physiological effect of morphine and heroin on the nervous system is not well understood, but it is generally thought that the drug affects synaptic transmission, probably by occupying sites on the receptor neuron so that transmission by normal biogenic amines is blocked.

Both the nitrogen and one or both hydroxyl groups are important, and so is the total molecular shape. That the phenolic OH group is important is shown by the fact that changing the structure there changes activity. Codeine with the methoxyl group in that position is much less active than morphine, while heroin with the acetyl group present there is about 20 times stronger than morphine. The fact that the total shape of the molecule is important is shown by the fact that small changes such as opening the nitrogen-containing ring or removing the single double bond markedly reduces activity.

Upon continued use, a tolerance develops for the opiate such that a large dose is required to produce the same effect of euphoria or analgesia. This tolerance disappears after the drug has been discontinued for a few days. When morphine or heroin are discontinued, a reversal of the effects on the body becomes evident. The effects of withdrawal are well-known. They have even been glamorized and overdone a bit in the press and in the movies and television. "Cold turkey" is nonetheless a distinctly unpleasant experience. The most common manifestations are a hyperirritability, chills and cramps, nausea, insomnia, and even delirium. The effects on the bowel is reversed leading to severe diarrhea. The withdrawal syndrome lasts from about 8 hr after the last dose of drug, up to 3 or 4 days. It is the first sign of this abstinence syndrome that leads the addict to seek another "fix".

What is there about heroin that leads to the phenomenon that we know as addiction? One obvious contributor is the beginnings of withdrawal and their unpleasantness. At the cellular level the development of tolerance suggests that the body tries to overcome the drug by developing more efficient receptors so that the effect of a dose is lessened. In any event, the addict finds the euphoria of the heroin experience so positive and the withdrawal effects so unpleasant that he is led to seek the drug continually. Indeed, the most notable fact about heroin addiction is that the chances of recovery are so small. While it is possible to overcome addiction, few addicts try, and many who do try then return to the drug after a time.

The death rate among heroin addicts is several times greater than that of comparable populations of nonaddicts. The effects of the drug itself are clearly traumatic on the body over the long term, but there is little to suggest a direct toxicity at normal dosages.

The opium poppy is cultivated for opium in some areas, notably Turkey and the nations of southeast Asia. The crude opium is processed illegally in clandestine laboratories in southern France and elsewhere. Here morphine is isolated and then acetylated. The pure heroin is then smuggled into the United States. Typically, heroin is "cut" by addition of an inert material, usually milk sugar. Street heroin will vary in heroin content, although there are obviously other factors involved in the cost such as the current supply/demand situation.

A heroin user usually injects the compound into a vein because this route enables him to experience the drug's effect in a matter of seconds. To do this the drug in powder form is first dissolved in water, with heat applied to speed the rate of solution, and the solution is administered by a hypodermic syringe. Repeated injections into a vein cause problems because the walls of the veins are weakened.

Drug users encounter several hazards. One of the most serious is the possibility of overdosage. Heroin overdosage can lead to death if the depressant effects lead to serious impairment of normal heartbeat and respiration.

A more usual hazard is from the deplorable conditions under which the addict usually operates. His hypodermic syringe is not likely to be sterile, and infection, especially hepatitis, is a frequent occurrence. Then there is the constant pressure of supplying the habit. Many an addict finds himself turning to crime to generate the necessary funds. This puts him in obvious jeopardy with the police and also with his supplier. It is not a life to envy. The price of the heroin euphoria is ultimately a very expensive one.

Treatment of the Addict

Given the problem of addiction, the next question is one of treatment. The standard treatment in the past has been to take the addict through the trauma of withdrawal, perhaps provide some psychiatric counseling, and then hope for the best. Synanon has used the approach of group therapy in which addicts counsel one another.

In recent years a new approach has been tried, one of minimizing the withdrawal effect by substituting another substance for heroin. The first such compound was methadone.

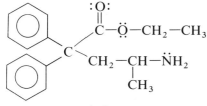

methadone

The difficulty with methadone is that one finds that the drug simply switches the addiction to methadone. Whether the addict is then better off is a debatable point.

New narcotic antagonists under development are being used in experimental treatment programs. Two of these are cyclazocine and naloxone. Their structures are

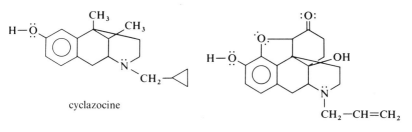

cyclazocine

naloxone

Note the similarity to morphine in structure with cyclazocine; the nitrogen is substituted by a cyclopropylmethyl group instead of a simple methyl, and the cyclohexene ring of morphine (see Fig. 11.10) is missing except for the two methyls. With naloxone the structure of the upper right-hand ring is modified, and the three-carbon unsaturated allyl group is present on the amino nitrogen.

Each of these antagonists is attracted to the morphine receptor centers in neurons. After taking the antagonist, a heroin or morphine injection produces no effect on an addict because the receptors are already occupied.

In early testing, disadvantages have been noted in the use of these antagonists. While cyclazocine relieves the desire for heroin, the patient tends to be tense and uneasy. With naloxone the patient is more relaxed, but the effect is shorter lasting than cyclazocine. So the ideal antagonist would be something which was itself nonaddictive with a long-lasting effect which left the patient with minimal discomfort. One idea to try would be to combine in one molecule some structural features of cyclazocine with others of naloxone. A new drug having the structure

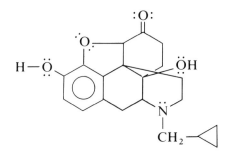

with some features of both molecules is under study as a possible narcotic antagonist.

About a dozen potential narcotic antagonists are presently in testing stages. All are similar structurally to those discussed above. While the ideal compound has not yet been determined, prospects are now very bright that this approach can soon offer a degree of relief to those addicts who would escape the tyranny of heroin.

SUGGESTED READING

1. Jerome H. Jaffe, "Narcotic Analgesics" and "Drug Addiction and Drug Abuse," in L. S. Goddman and Alfred Gilman, *The Pharmacological Basis of Therapeutics,* The Macmillan Company, New York, N.Y., 1965. This reference discusses the addiction problem of the opiates from the pharmacological standpoint.

2. G. M. Dyson and Percy May, *May's Chemistry of Synthetic Drugs*, Longman's, Green & Co., Ltd., London, 1959, pp. 142–170. These pages discuss chemical structure and synthesis of various opium alkaloids.

3. Paul H. Blachly, ed., *Drug Abuse—Data and Debate*, Charles C. Thomas, Publisher, Springfield, Ill., 1970. This surveys various drugs from the socio-political and medical standpoints.

4. *Drug Addiction: Crime or Disease?* Report of a Joint Committee of the American Bar Association and the American Medical Association on Narcotic Drugs, Indiana University Press, Bloomington, Ind., 1963.

5. Stanley Einstein, ed., *Methadone Maintenance*, Marcel Dekker, Inc., New York, N.Y., 1970.

6. A. L. Hammond, "Narcotic Antagonists: New Methods to Treat Heroin Addiction," *Science*, **173**, 503–506 (1971). This article discusses the present state of knowledge relative to cyclazocine and naloxone.

EPILOGUE

This section has been a survey of what chemical and physiological information we have on those drugs that are of great social concern today. There are a few public impressions that need to be corrected. One is the tendency to overgeneralize about "drugs." We need to be clear about the facts that some drugs are addictive (alcohol, barbiturates), that some are not (amphetamines, marijuana), that some can cause permanent damage to man (LSD, heroin), and that others may be nearly harmless (marijuana and alcohol in moderation). There is, in fact, very little that the classes of compounds have in common, except that they affect the central nervous system by altering perception and mood and that they are all illegal or tightly controlled. Whether the legal restraints on some should be loosened or tightened is an important political question that must be decided in the United States and elsewhere.

Finally, the point of insufficient information has been made again and again. In order to decide how to deal with these substances and new ones that are emerging, we must know more about the chemistry of the nervous system and about the impact made by drugs at the molecular level. Some argue against research on grounds that it can be manipulated to show whatever conclusions that some private interest favors. While this has happened in the past, these critics do not understand science or scientists. Any scientist true to his calling (and that is most of us) will not permit the results of his work to be so manipulated. Given enough manpower and equipment, the chemist, physiologist, and pharmacologist possess the ability and capacity to obtain the answers society needs to deal competently with these questions.

PART IV

SCIENCE — THE ETHICAL AND POLITICAL DILEMMA

12

The Scientist — The Ethical and Political Dilemma

> For to know more and more about less and less
> is in the end simply to know less and less.
>
> *Lewis Mumford*
> *Distinguished American Scholar,*
> The Pentagon of Power, *p. 181*
>
> Science is humanism.
>
> *G. N. Lewis (1875–1946)*
> *Distinguished American Chemist*

Chemical Orientation In this chapter we shall review concepts of nuclear chemistry and you may wish to review sections of your text related to that topic. Also, one could review those sections of Chapter 1 having to do with this subject (pp. 39–46). Concepts of hydrogen bonding as they relate to biologically important molecules appear in this chapter and should be reviewed. If your text presents discussions of protein structure and the nature of enzymes and their activity, you should read them carefully.

Finally, your text may contain a section devoted to historical development of chemistry or an outline of the philosophy and methods of the chemist. This is frequently the first section of a text.

SCIENCE AND SOCIAL IMPACT

The strength of science is the scientific method. With its continuing interaction between experimental fact and theoretical interpretation and its insistence that a scientist's finding be subjected to rigorous judgement by his peers, the scientific method has been the factor which has led to the explosion of knowledge and technological advance which characterizes the nineteenth and twentieth centuries. On the other hand, there are difficulties in how science and scientists operate. One is that in establishing judgement by the peer group, little account is given to those outside. Scientists tend to develop a language of their own that is unintelligible to a layman or even to a scientist in a different field. In the field of education, scientists have given very great attention indeed to the teaching of science to potential scientists but relatively little to the teaching of science to the nonscientist. Yet, communication between scientist and nonscientists, which such education would foster, is important.

Because of this communication barrier, difficulties develop in the decision-making process. In order for scientific research to proceed, money must be provided for the requisite salaries and equipment. This money is provided by government subsidy or by private corporations. The power of the purse is the ultimate power, and so those who make these decisions hold the power of deciding where research effort will be directed. They control what will happen to science to a degree that scientists do not. We can expend great efforts at conquering cancer, developing new missiles, or going to the moon. The scientific community, or parts of it, may flourish or wither according to these decisions. The problem is that the decision-makers are not generally scientists, so that they do not themselves possess the competence to judge the scientific evidence. The usual practice is for them to rely on scientific experts to explain the scientific work. The other horn of the dilemma is that scientists are not particularly competent to evaluate the social impact or potential of a given research problem, except in terms of the science itself.

In this section we explore some areas of this confrontation of scientist and nonscientist decision-maker: in particular, we shall examine the interaction between the scientist and politician. We shall look at the decision-making structure of science and for perspective we shall assess the impact or potential impact of specific scientific developments both from the present and from the past. Four case studies are examined. Two examine past technological advances whose impacts can be assessed. The other two are in differing stages of development and the dilemma of future impacts of impending social change is only beginning to emerge.

CASE STUDY 1: "MANHATTAN DISTRICT, U.S. CORPS OF ENGINEERS"

In the spring of 1939, a German physicist Otto Hahn made known the results of an experiment in which uranium was bombarded by slow neutrons. Hahn had detected barium as a product. That element was produced by the fission of the isotope U^{235} as we showed on p. 43. Two facts were immediately clear to physicists all over the world. One was that neutrons must have been produced in the reaction so that a chain reaction could be created, neutrons occurring as both reactants and products. The other was that large quantities of energy were released during the reaction. In the fission reaction, 1 mole of U^{235} releases 4.7×10^9 kcal, more than a millionfold greater energy release than conventional exothermic chemical reactions. The potential for uses of this energy was immediately apparent in the scientific community. The fission reaction could be used to generate energy for peaceful or military activity. In 1939 the prospects for war were intensifying, and this led to concentration on the latter possibility.

It was obvious that Hitler was preparing to wage war. The Munich Conference of the previous fall and the subsequent occupation of Czechoslovakia had served to assuage but not eliminate his aggressive appetites, and already the shadow of confrontation with the Allies over Poland was developing. A number of scientists had themselves fallen victim to Fascism in Europe, including Albert Einstein and Enrico Fermi. Some found the environment too oppressive for openness and freedom to pursue whatever research they wished, and they emigrated to the United States. Jewish scientists fled to avoid Hitler's anti-Semitism. These scientists were destined to play key roles in the development of the atomic bomb.

In spite of the potential of the fission reaction there was an enormous technical problem. It was that the active isotope U^{235} constituted only 0.7% of naturally occurring uranium. Most of the rest is U^{238}, a stable isotope. The amount of U^{235} that was present was so small that the neutrons released in natural uranium would not sustain the chain reaction. The chances of it striking another U^{235} were not good enough, and the chain reaction could not continue. In order to generate a bomb, it was necessary to release a large amount of energy all at once, and this required that U^{235} be present in higher concentration. The most important objective in building the atom bomb was to concentrate U^{235} to a *critical mass** large enough to sustain the fission chain reaction.

Otto Hahn's work had been done at the Kaiser Wilhelm Institüt in Berlin, and United States scientists knew that the German physicists and chemists were capable of both defining and solving the problem. The possibility that Germany might possess a nuclear fission device and could conceivably win the impending war thereby led the concerned scientists, particularly Dr. Leo Szilard, to seek an initiation of research on nuclear fission from President Franklin Roosevelt. A letter transmitted to the President under the signature of Einstein led to the establishment of a "Uranium Committee" in the fall of 1939.

There were three experimental problems to be solved. The first was the requirement of experimental demonstration that a chain reaction was occurring. Experiments at the University of Chicago using an atomic pile in which U^{238} was bombarded with neutrons producing U^{239} that decayed by β-emission to plutonium-239, as shown on p. 44.

The chain reaction was moderated by use of graphite rods that absorbed sufficient neutrons to keep the fission reaction under control. The sustained chain reaction was successfully carried out on December 2, 1942, under the direction of Fermi and the United States physicist Arthur Compton. The advantage of a plutonium fission approach was the availability of Pu^{239} from U^{238}, the dominant isotope in natural uranium. The disadvantage was the enormous output of power required, 500 to 1,500 kw, to produce 1 g/day. Plutonium production on a large scale was undertaken in hopes of producing a plutonium bomb device.

At the same time, efforts were underway to effect the enrichment of U^{235} from natural uranium. The problem with separation of isotopes was that the chemical and physical properties are nearly identical, since the electronic configuration and atomic size are the same. There are very slight differences in some properties of U^{235} and U^{238} atoms because of their differences in mass. For instance, uranium hexafluoride (UF_6) is a gas at slightly above room temperature, even though the molecular weight is high. Since $U^{235}F_6$ has a slightly lower mass than $U^{238}F_6$, some of its properties that depend on mass will be different. One of these is the average velocity of the gas molecules moving in space at a given temperature. This property led to a separation technique known as *gaseous diffusion* where the gas diffusing through barriers would lead to gradual enrichment of $U^{235}F_6$ among those molecules that were diffusing fastest. By separating out the fastest moving fraction, a U^{235}-enriched sample of UF_6 could be isolated.

A second separation approach attempted use of the cyclotron.* In this method a gas is ionized and the gaseous ions are accelerated through a magnetic field. In this situation the moving ion describes an arc as it is deflected in the magnetic field. Particles with smaller mass, the U^{235}-containing ions, are deflected more because their lower mass gives them less momentum. Thus, there will be two beams of particles emerging from the magnetic field chamber and they can be separately collected. One beam gives the U^{235} isotope; the other, the U^{238}.

Operations for the study of gaseous diffusion were established under the leadership of Harold Urey of Columbia University. Ernest Lawrence of the University of California directed the cyclotron separation effort. The latter method proved less successful, although it did effect the separation of quantities of U^{235}. The gas diffusion method was successful and led to production of U^{235} at plants constructed at Oak Ridge, Tennessee, for the purpose. The total effort was given the code name "Manhattan Project", and rigorous security was imposed. Thousands of scientists were brought into the project in laboratories across the nation. Few knew the overall objectives of the research. Most members of Congress appropriated large sums to the project knowing nothing of its purpose.

Once the problem of obtaining the raw material had been solved, the next step was to build the bomb. Working against the pressure of the war and the possibility that Germany might be progressing on a development program of its own, scientists assembled in a remote desert area of New Mexico, on a mesa called Los Alamos. The director was the theoretical physicist J. Robert Oppenheimer, a long-time colleague of Lawrence at the University of California. Los Alamos began operation in the spring of 1943.

One of the problems to be overcome was predetonation. The most desirable thing was to assemble a mass of sufficient size to produce the desired explosive intensity. Assembling the U^{235} was complicated by the fact that once the critical mass had been achieved, the sustained chain reaction would produce too much heat and pressure to permit the larger quantity required for a massive explosion. Thus, a small explosion, but not the larger one, would result. The problem was essentially a mechanical one of firing part of the uranium as a projectile into the rest as target so rapidly the predetonation was avoided.

The first nuclear explosion occurred in the early morning hours of July 16, 1945. By that time Germany had already surrendered without having successfully developed a nuclear device. The Hiroshima and Nagasaki bombings ended the war against Japan. The decision to use those bombs had been made by President Harry Truman.

From there the commitment to nuclear weaponry became inevitable because of the competition of the Cold War, and scientists in a number of nations have since been deeply embroiled in nuclear research leading to the nuclear fusion bomb and the sophisticated missile delivery systems. As well, they have been engaged in research on peacetime uses of nuclear energy. A few scientists were repelled by the idea of the bomb, but most saw the situation as one in which development of the bomb before Germany was necessary for national survival. When the destructiveness of the device was seen, some regretted their participation, feeling that humanity would have been better served had the secret of nuclear energy not been revealed to man.

Many nuclear physicists took the position that they and their colleagues had been concerned too little with the impact of their work until too late. At any rate, though, the situation that we face is perilous; optimists see a sobering effect being felt by the leaders of the Great Powers as a result of their ability to destroy one another and the rest of mankind. The adoption of the Nuclear Test Ban Treaty by the United States and the Soviet Union in 1963 has led to diminution of the atmospheric content of radioactive nuclides from weapons testing. The SALT agreements of 1972 led to an arrangement whereby the United States and the Soviet Union agreed to a preliminary cutback in missile levels. Perhaps the future will see further political agreements leading to the end of the nuclear balance of terror.

SUGGESTED READING

1. H. D. Smyth, *Atomic Energy for Military Purposes: The Official Report of the Development of the Atomic Bomb Under the Auspices of the United States Government, 1940–1945*, Princeton University Press, Princeton, N.J., 1946. This is the official account of the Manhattan Project and developments leading to it.

2. Ralph E. Lapp, *Atoms and People*, Harper & Brothers, New York, N.Y., 1956.

3. Nuel P. Davis, *Lawrence and Oppenheimer*, Simon and Schuster, New York, N.Y., 1968. This last reference is a highly personalized account centering around the two dominant figures in United States physics during mid-twentieth century, in terms of their contribution to the Manhattan Project.†

CASE STUDY 2: CBW — PUBLIC HEALTH IN REVERSE

After its beginning months, World War I quickly settled into a stalemate in the trenches on the Western Front. Each side was able to prevent the other from advancing because there was no adequate offensive strategy to counteract the fire power that each side was able to generate with automatic weapons from protected positions in the trenches.

In seeking an offensive strategy to attack trench positions, Germany turned to one of her great strengths, the advanced technology of her sophisticated chemical industry, to produce toxic gases. The strategic advantage of a gas was that it could penetrate through without resistance. To its disadvantage was the difficulty of wind changes in pinpointing the site for a gas attack and the fact that the attackers were liable to the effects of the gas as well as the defenders.

The first gas attacks of major proportions by the Germans occurred at Ypres against French and Canadian troops in 1915. The attack was successful, and Allied lines were broken. The toxic agent used was chlorine, Cl_2, which is a reactive oxidizing agent. In reacting with tissues of the respiratory tract, it exerts its toxic effect by so irritating the mucous lining that the victim is soon disabled and dies of asphyxiation. Two things happened quickly: The Allies responded with gas attacks of their own, and a proper defense, the gas mask, was developed. As a result, gas gave no particular offensive advantage, although it was used frequently by both sides.

Nevertheless, several toxic gases were developed and used during the conflict. Among these were phosgene,

† "There floated through my mind a line from the Bhagavad-Gita in which Krishna is trying to persuade the Prince that he should do his duty: 'I am become death, the shatterer of worlds.' I think we all had this feeling more or less." J. Robert Oppenheimer concerning his reaction to the Los Alamos blast. From *Lawrence and Oppenheimer*, p. 239.

which is also a lung irritant, a more severe one than Cl_2. It attacks free hydroxyl (OH) and amino (NH_2) groups producing HCl and heat, which are responsible for the irritant effect. The reaction occurs with water and with hydroxyl and amino groups of molecules in the tissue.

$$R-\ddot{\underset{\cdot\cdot}{O}}-H + :\ddot{\underset{\cdot\cdot}{C}}l-\underset{\underset{:O:}{\|}}{C}-\ddot{\underset{\cdot\cdot}{C}}l: \longrightarrow R-\ddot{\underset{\cdot\cdot}{O}}-\underset{\underset{:O:}{\|}}{C}-\ddot{\underset{\cdot\cdot}{C}}l: + HCl + heat$$

(This chlorine reacts as did the first)

Mustard gas is a vesicant or skin inflammatory agent, and its effect is to cause severe slow-healing blistering on any part of the body that it contacts.

$$:\ddot{\underset{\cdot\cdot}{C}}l-CH_2-CH_2-\ddot{\underset{\cdot\cdot}{S}}-CH_2-CH_2-\ddot{\underset{\cdot\cdot}{C}}l:$$

mustard gas

Toxic gases were not used during World War II. The reasons for this may have been the Geneva Convention of 1925, which forbade using lethal gases. Most combatants were signatories. A cynic may say that the reason was the ineffectiveness of gas during World War I or that no one saw themselves in a position to gain by its use. Since World War II, the use of toxic gas against military or civilians has not occurred in other wars, but other chemical weapons have been developed. In particular, the extensive use of napalm as an incendiary and of herbicides as defoliants by the United States in Vietnam in the late 1960's should be noted. As well, in a few instances, the use of non-lethal lachrymators or tear gases has been established.

Technological developments in the United States, the Soviet Union, and other countries has led to a number of other chemical agents. The most toxic and feared are the nerve gases, which are structurally related to phosphoric acid. Take, for example, the compound known by the code symbol, GA.

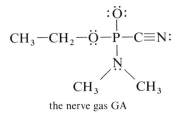

the nerve gas GA

Most of these compounds act as antiacetylcholinesterases; by preventing the hydrolysis of the biogenic amine acetylcholine (p. 254), they disrupt coordination in the autonomic nervous system and in the muscles, causing tremors, convulsions, and death. GA is about 10 times as toxic as phosgene. Other nerve gases are still more toxic. Thus, their activity is much like that of the organophosphate pesticides (p. 142).

Also important in the military arsenal are agents that incapacitate but do not kill. Lachrymators or tear gases have been used routinely in civilian police work and at times by the military. One of the chief tear gas agents in use is phenacyl chloride, also known as CN:

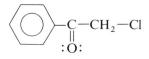

Another agent, CS, produces both tearing and nausea. Even LSD and other hallucinogenic drugs have been considered as potential chemical agents.

On the biological side, the United States and other great powers have developed the capability to deliver bacterial or viral disease agents, such as anthrax, tularemia (rabbit fever), Rocky Mountain spotted fever, psittacosis (parrot fever), and plague. The characteristic of these diseases is that they are highly discomforting, causing weakness and pain. Mortality tends to be high. Thus, they debilitate as well as kill. Production of the toxins required is carried out usually by standard techniques of microbiology by injecting the toxin into a culture such as a fertile egg in which it will multiply rapidly. Research programs on storage, delivery systems, and the development of methods to combat them are important aspects of the military CBW programs.

Information about CBW is difficult to obtain because of military secrecy. The level of spending for military CBW has been about 150 million dollars/ year. In late 1969 President Nixon announced that the United States would not use biological warfare methods under any circumstances, and he reaffirmed a previous position taken by other Presidents—that the United States would not use chemical agents in warfare unless they were used first by the enemy. In accord with this decision, research and production of biological agents has been curtailed, and research programs at Pine Bluff, Arkansas; Fort Detrick, Maryland; and the Dugway Proving Grounds, Utah have been reduced. Large stocks of nerve gas have been disposed of by sinking them at sea. Production and research in chemical agents continue, however.

Many in the chemical warfare service argue that the use of chemical agents in warfare is more humane in some circumstances than conventional methods. They base this on the fact that there are chemical agents that disable temporarily but do not kill. The dominant response to CBW by most people in all nations seems to be one of revulsion because of their bizarre effects and because of the potential to bring death to civilians as well as the military combatants.

Those scientists working in the CBW area do not participate fully in the free interchange of ideas that is the keystone of sciences. This is because most of their work is secret and is therefore not subject to evaluation by their scientific peers generally. The moral dilemma that these scientists face is a real one. Under the present world political situation, the United States takes the position that it must be able to respond to a chemical attack by replying in kind. The same argument—of preventing attack by maintaining a retaliative strength—is used for maintaining a nuclear arsenal. The scientist who works in this field must resolve this view with the possibility that his work may lead to suffering and death of many human beings.

SUGGESTED READING

1. Michael McClintock, *et al., Environmental Effects of Weapons Technology,* Scientist Institute for Public Information Workbook, SIPI, 30 East 68th St., New York, N.Y. 10021.
2. Elinor Langer, "Chemical and Biological Warfare, I. The Research Program," *Science,* **155,** 174–176, 178–179 (1967); II. "The Weapons and Policies," *Science,* **155,** 299–303 (1967).

CASE STUDY 3: THE MOLECULAR BASIS OF GENETICS

What We Have Here is Successful Communication

The single most important scientific advance since World War II has probably been in our understanding of the structure and behavior of genes on the molecular level. By this knowledge we now understand fairly well some of the most fundamental properties of biology: Cell division, protein synthesis, reproduction, and mutation have all become better understood. At the same time we can begin to see possibilities that may be hopeful or ominous for using this information. As scientific knowledge translates into technological advance, it becomes conceivable that man could restrain the aging process and cure genetic diseases such as hemophilia and sickle cell anemia. The possibility of designing human beings to meet certain physical, chemical, or biological specifications is more remote, but "conceivable". Most important, the advances in molecular genetics will enable us to understand and deal more effectively with the diseases we call cancer.

In this case study, we deal with the question of the impact of the new genetics on society and what the role of the scientist may be in dealing with the potential disruptions that it may bring on social institutions and moral concepts. Here we deal with a possible future impact of scientific advance on society that may greatly change not only society but man himself.

Studies of cells of all plants and animals reveal a core area, called the *nucleus*, that is clearly distinct from the rest of the cell. The nucleus is necessary in the function of the cell and its division causes the cell to divide. Within the nucleus, distinctive bands of matter called *chromosomes* are observed. In the human there are 46 chromosomes. During cell division the chromosomes appear to divide along their longitudinal axis, so that the key to cell division, and cell function as well, exists in the chromosomes. A *gene* is a piece of inherited information or a single inherited trait. A given chromosome will contain many different pieces of information. Much of the gene content of a given cell will be the same for all individuals in a species. Thus, all humans undergo similar growth and development patterns, have nearly the same enzymes in their cells, and generally resemble one another more than they resemble bullfrogs or grizzly bears. At the same time, there is provision for individual differences that are also inherited from one's parents. Thus, the hair color, body conformation, and many other features will be different. Provision is also made for *mutation*, or changes in genetic information that occur for some reason. Usually these are disadvantageous, but sometimes they can be benign or even beneficial.

The New Genetics

The question that we have begun to unravel is the molecular structure of the gene and chromosome. The key molecule for all plants and animals, except the simplest viruses, is deoxyribonucleic acid, known as DNA.

The DNA molecule is made up of three structural subunits. They are

1. Phosphate groups:

2. The sugar deoxyribose:

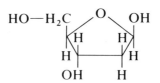

related to ribose

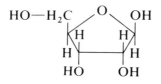

except that an oxygen is missing.

3. Heterocyclic amines called *nucleotides*. There are four of these in DNA (Fig. 12.1). The subunits fit together such that the phosphate groups form a polymeric bridge structure by forming phosphate ester linkages with the deoxyribose molecules.

1. Adenine   2. Guanine

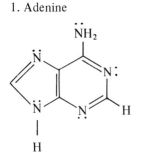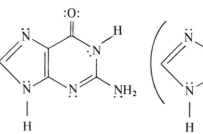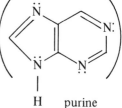

purine

3. Cytosine   4. Thymine

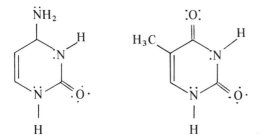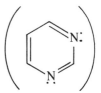

pyrimidine

FIG. 12.1 Nucleotides of DNA. Note that the first two share a common ring structure. For that reason they are called the *purine* nucleotides. Cytosine and thymine likewise share the common ring structure of pyrimidine.

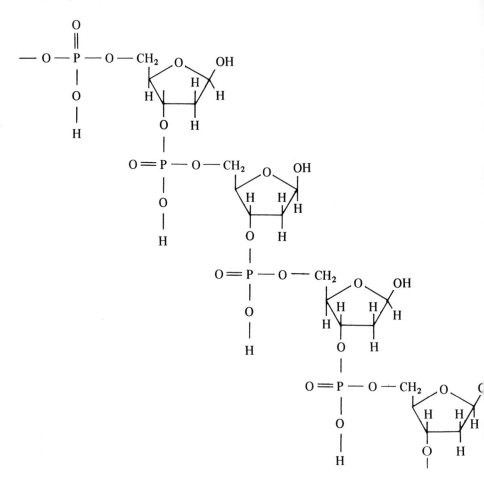

FIG. 12.2 Alternating bridge of DNA backbone comprised of deoxyribose and phosphate units.

There are three hydroxyl groups on deoxyribose, at carbons 1, 3 and 5.

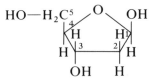

By forming phosphate esters through the groups at carbons 3 and 5, a chain of alternating sugars and phosphates is achieved. This forms the so-called "backbone" of DNA. The nucleotides attach to the backbone through

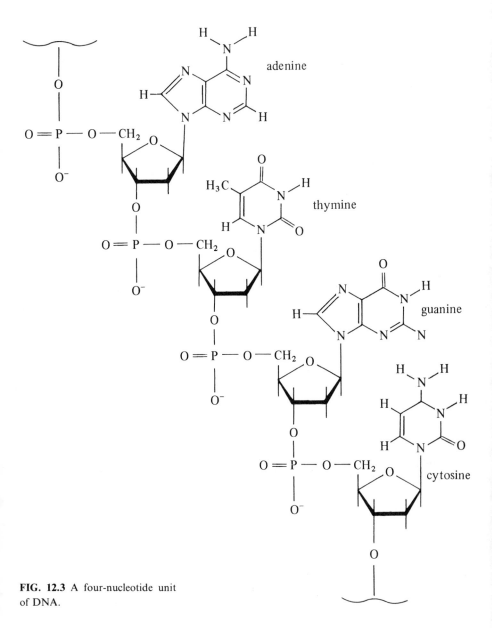

FIG. 12.3 A four-nucleotide unit of DNA.

amine groups to the deoxyribose at the carbon−1 hydroxyl group. A segment of four units would thus have the structure in Fig. 12.3. The DNA chain is very long, consisting of many millions of nucleotide units. The

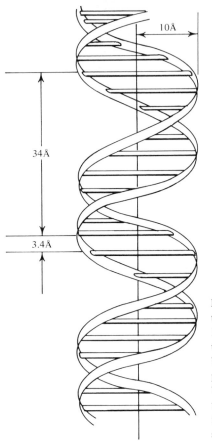

FIG. 12.4 The double helical structure of DNA. Two bands are the
DNA backbone and the connections between them are generated
by interactions between nucleotide
pairs. The helical radius, distance
between units of the backbone, and
the length of the helix through one
full turn are all indicated in Ångstrom units.

important structural factor in DNA that makes genes different is the order in
which the nucleotides occur.

The DNA structure is still incomplete. We have only told half the story.
There are two chains that go together to form the molecule. The two chains
are intertwined in a structure that has become known universally as the
"double helix" (Fig. 12.4).

The helix forms by a pairing of the nucleotides. Each of the four has a
partner. Thymine and adenine are paired, as are guanine and cytosine.
The bases are linked together by hydrogen bonds. Whenever adenine
appears on one chain, thymine will be present on the other, and similarly with
guanine and cytosine.

Now we can reach a simple understanding of how the molecule functions
to transmit genetic information. We can envisage how DNA functions in cell
division. First, the double helix is unraveled by breaking the hydrogen bonds.

The thymine-adenine linkage:

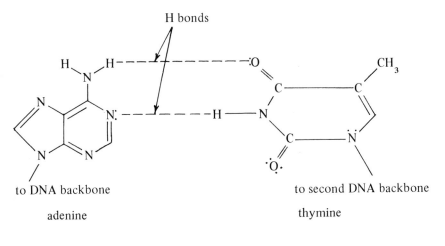

The cytosine-guanine linkage:

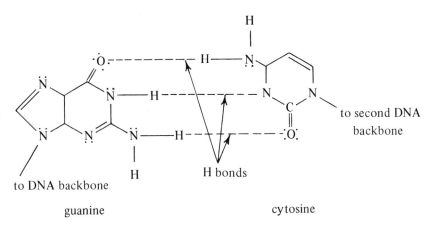

FIG. 12.5 Interaction of nucleotides in double helix.

Then each half, with the help of enzymes, synthesizes its opposite member by forming a new chain of partner nucleotides. At the end we have two new DNA molecules identical to the one with which we started. Each is capable of governing a separate cell, and the old cell divides to form two new ones.

The process of fertilization may be understood also. Each sperm and ovum contains half the DNA information. At fertilization, the two chains couple to form the new organism DNA, with traits of both parents being incorporated into the progeny DNA.

The most important function of the cell is to synthesize all the enzymes that perform the chemical reactions required by the organism. DNA is involved in this operation—but not directly. Rather it is involved in the synthesis of a second nucleic acid polymer, which differs in two important respects from DNA. First, ribose is used as the sugar in the backbone instead of deoxyribose (see p. 311). Then there is one change in the nucleotides, as thymine is replaced by a very similar pyrimidine, uracil.

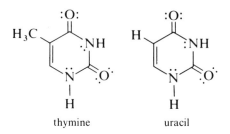

thymine   uracil

With these two modifications, we have a slightly different polymeric structure, ribonucleic acid, or RNA. DNA contains the code by which RNA synthesis occurs. Again with the assistance of enzymes, a portion of the DNA helix uncoils and RNA is synthesized by matching nucleotides under the catalytic action of an enzyme, RNA-polymerase. The new molecule migrates from the nucleus to another portion of the cell called a *ribosome*. Here the synthesis of protein molecules takes place under the direction of RNA.

The important thing in protein synthesis is the amino acid sequence, and the function of RNA is to communicate this sequence. The RNA has built into it a code that is communicated to the synthesis. The code, handed down from DNA, is the nucleotide sequence. For each amino acid, a three-nucleotide unit is required. Thus, a sequence of nucleotides dictates a sequence of amino acids. The translation from nucleotide language to amino acid language is known as the *genetic code*.

In the synthesis of the peptide chain each sequence of nucleotides, in pairs of three, calls forth a peptide. Since there are 4^3 or 64 possible different sequences and only 20 amino acids, there is some duplication as one can see from Table 12.1. Three nucleotide triplets, or *codons*, are used to punctuate; that is, they signal the termination or initiation of chains. In Table 12.1 they are indicated by the asterisks. We can see how the code works from an example. A nucleotide sequence, UUG, CAU, ACC, GGU, UAA would cause the synthesis of an amino acid chain leucine, histidine, threonine, glycine with termination of the chain indicated by the last codon.

TABLE 12.1
THE GENETIC CODE[1]

		Second Nucleotide				
		U	C	A	G	
F		Phenylalinine	Serine	Tyrosine	Cysteine	U	T
i		Phenylalanine	Serine	Tyrosine	Cysteine	C	h
r	U	Leucine	Serine	*	*	A	i
s		Leucine	Serine	*	Tryptamine	G	r
t							d
		Leucine	Proline	Histidine	Arginine	U	
N		Leucine	Proline	Histidine	Arginine	C	N
u	C	Leucine	Proline	Glutamine	Arginine	A	u
c		Leucine	Proline	Glutamine	Arginine	G	c
l							l
e		Isoleucine	Threonine	Asparagine	Serine	U	e
o		Isoleucine	Threonine	Asparagine	Serine	C	o
t	A	Isoleucine	Threonine	Lysine	Arginine	A	t
i		Methionine	Threonine	Lysine	Arginine	G	i
d							d
e		Valine	Alanine	Aspartic acid	Glycine	U	e
	G	Valine	Alanine	Aspartic acid	Glycine	C	
		Valine	Alanine	Glutamic acid	Glycine	A	
		Valine	Alanine	Glutamic acid	Glycine	G	

[1] Taken from Phillip Handler, ed., *Biology and the Future of Man*, Oxford University Press, New York, N.Y., 1970, p. 42.

There is in this arrangement a threefold interdependence among DNA, RNA, and proteins. As DNA generates RNA, and RNA triggers protein synthesis, it is still required that enzymes initiate the synthesis of DNA in cell division, so the cycle is complete. The enzyme, RNA-dependent DNA-polymerase, is required for DNA synthesis.

The keys to life are the enzymes, which do almost all the cells' chemical work, catalyzing reactions that provide energy, form new tissue, and so on. The important thing to realize is that enzyme activity is tied to the sequence of amino acids. Enzymes act usually by providing a surface on which a reaction occurs. Generally a reactant molecule is absorbed to form an enzyme-reactant complex. The reaction occurs on the enzyme, forming product molecules that then leave the surface. The shape depends on how the long chain of amino acids folds and winds around on itself, and this depends on the sequence. The shapes of enzymes are known for only a few enzymes and the reactivity of an enzyme as a function of enzyme shape remains a field of importance to be explored. Determining the structure* of an enzyme having, say 300 amino acids in sequence, is a difficult task, almost impossible with today's technology. The development of new analytical methods, however, especially sophisticated X-ray analysis connected with computers, leads one to expect that the structure determination of enzymes will soon be easier.

So we see that a healthy cell requires active enzymes doing their chemical work; that this depends on the correct sequence of amino acids; and that this depends on the correct nucleotide sequence of RNA, which in turn depends on the corresponding DNA sequence. When one considers that the human cells require thousands of different enzymes constructed of hundreds or thousands of amino acid sequences so that the number of nucleotides required is astronomical, a huge amount of chemical detective work becomes necessary.

The miracle of life is that all this has happened so flawlessly through cell divisions and reproductive acts, adjusting to changing environment and eventually evolving new species, on an increasing level of complexity through eons of time. The present state of genetics is that while we cannot understand all of it, we know enough to realize that man has far indeed to go to match nature's chemical ingenuity.

Implications of the New Genetics: Theory of Cancer

With our understanding of the genetic code, a general interpretation of some biological phenomena is possible. *Mutation* is a change in nucleotide sequence leading to a difference in enzymes, which in turn causes a change in the organism. If the change is favorable, the organism is better equipped to survive and reproduce, which will tend to pass the mutation to its offspring. If the mutation is unfavorable, individual survival is less likely and reproduction and inheritance of the unfavorable mutation is less favored. Favorable mutation will tend to lead to changes in a species and this means evolution, as natural selection operates. *Growth* is a process in which cell division is occurring faster than old cells are being replaced. Growth and suppression of growth require that DNA replication be turned on and turned off properly by actions of hormones. *Aging* involves, in part, a gradual breakdown over time in the efficiency of information transferral so that more and more ineffective or inefficient cells build up, leading finally to the breakdown of a critical function, causing death.

Cancer is an aberration of normal genetic information such that growth is unrestrained. The key to stopping cell division at the proper time fails to function. The very word cancer is spoken with dread and may not even be uttered in certain circumstances. Visions of pain and lingering and inevitable death creep in, and too many of us have had the experience of loss. Three hundred thousand Americans die each year of the affliction. Cancer is really many diseases and, as such, the problem is multiple. Indeed, the only common feature among them is the unrestrained aberrant cell division. But because of this common feature, it is worthwhile to inquire about cancer from the point of view of molecular genetics.

In particular it may be worthwhile to consider the question that cancer may be caused by a virus. In the molecular world and the biological world the virus is in an interesting position, for it represents the boundary between lifeless molecules and living cells. The virus is in fact a DNA molecule covered by a protein sheath. In order for the DNA to function, it must find itself a cell. Then it somehow incorporates itself into the cell DNA in a sort of cytological hijacking. Now the cell functions at the command of the viral DNA and replication of virus DNA and RNA take place. This way the virus reproduces and cell division takes place.

The body's response mechanism is the same as that for any other foreign invader. It sends out its army of defenders, the antibodies* who remove the foreign cells. For some reason the body may not possess the capacity to synthesize the antibodies required to stop the cancer, hence their indiscriminate growth.

In addition to the virus theory, there are other ideas relating to how the disease begins. Known carcinogenic agents, chemical substances known to cause the development of tumors, exist. The concern about correlations of cigarette smoking and cancer arises from the presence of small quantities of carcinogenic hydrocarbons in cigarette smoke, or for that matter any smoke arising from combustion of organic matter. Radiation, such as X rays or gamma rays, is another source of cancer. A final possible cause is that there is simply a genetic error which is not corrected by the mechanisms in the cell which are charged with correcting genetic misinformation.

The present approach to treating the disease is oriented toward early detection and elimination of the cancerous tissue. Surgery is the method of choice frequently; the others are to kill the cells either with a chemical or with radiation. The problem with either approach is that unless every diseased cell is removed, the tumor may reemerge after a few months or years. These methods also can lead to the death of normal cells. If too much of this happens, the cure will be worse than the disease.

The nation has decided to apply its research talents and energies to an effort to improve our defenses against cancer. A large sum of money has been approved for cancer research. Investigations will go forward in all areas, in chemotherapy, radiation therapy, surgery, and investigations of the role of virus in the cancer phenomenon.

The approach is envisaged to be in the nature of a crash program, not unlike the Manhattan Project, designed to get answers fast—but there will be a cost. The cost will be that, not understanding the fundamental chemistry of cancer, we are less able to predict intelligently in advance what are likely to be effective solutions. As a result, a shotgun approach may enable us to find good methods of treatment, in the midst of many approaches that will not be fruitful. At the same time, it is important to continue to study the fundamental cell biology and biochemistry involved in cancer cell growth and in the takeover of host cells by viruses.

Applications of the New Genetics

Beyond the cancer question, the knowledge of molecular genetics that we have gained in the last 2 decades points toward other applications of the new genetics, some hopeful and others foreboding. Consider the question of genetic diseases. The failure to synthesize correctly a single enzyme may lead to a serious disease problem. If the problem is sufficiently severe, the unfortunate fetus will probably not live to full term, which is how natural selection operates to remove genetic problems by nipping them in the bud. There are some well-known genetic diseases that impair but do not kill, however. Among these are forms of mental retardation, including mongolism and the disease phenylketonuria, which causes mental retardation by improper metabolism of the amino acid phenylalanine in key stages of brain development during infancy. Some forms of diabetes, the disease of insufficient pancreatic production of insulin, an important enzyme in carbohydrate metabolism, is sometimes inherited. Hemophilia, the disease of the nineteenth-century European nobility, is caused by failure of blood clotting mechanisms. Sickle cell anemia, affecting the black race mainly, is caused by a decrease in efficiency of oxygen transport by red blood cells due to a genetic phenomenon.

The first problem in research in the genetic diseases is to identify the genetic site. It is possible that a single nucleotide is misplaced, which could cause one error in amino acid synthesis. For example, suppose an RNA sequence was CAG instead of CAU because a guanine was erroneously placed where a uracil nucleotide should be. The amino acid placed in an enzyme being synthesized would then have a glutamine instead of a tyrosine unit in that position of the enzyme (Table 12.1). That might be of no importance, or it might be critical, depending on whether the change altered the key active sites of the enzyme or not. One single such error could cause an enzyme to lose its catalytic ability. Then a single reaction, important in a series of metabolic events, fails causing a whole series of untold effects, not unlike the nursery rhyme in which the loss of a nail, hence a shoe, hence a horse, etc., leads in the end to the loss of the kingdom.

If the error at the site of the gene were known, then perhaps one could perform an act of genetic surgery, designing a way of removing the offending DNA segment and replacing it with the correct one. Perhaps this could be done with a chemical, but chemical agents are likely to be too indiscriminate. It is more likely that some sort of a beneficial virus could be found or invented to do the job.

Experiments leading to man-induced changes like this in DNA lead to intriguing, some say ominous, possibilities. In one experiment an unfertilized tadpole egg with only half its DNA had that DNA substituted for by the DNA of an intestinal cell of another tadpole. With a complete genetic packet, the egg developed as it would if it were fertilized by a spermatozoon, producing an identical twin of the second tadpole. The technique is called *cloning*, and it has been demonstrated several times in different species. It is conceivable to think of cloning a man by, for example, taking a cell from a great musician, injecting it into an unfertilized human egg, implanting this into a uterus (or perhaps not even that) and producing a genetic twin of the donor.

Then genetic planning becomes foreseeable as the next step. If one could pinpoint the site where certain genetic information was located, this information could be changed or specified according to a preplanned destiny for the developing embryo.

Let us now look at the ethical problems for the scientist and for society. One problem in this area is with us now. It is the question of the right of an individual to reproduce, even though there is a good chance that the offspring will contain genetic defects. Conventional genetics is sometimes sufficiently advanced to tell us when this is the case. By permitting reproduction and inheritance of genetic defects, we appear to weaken the genetic pool, the available information stored collectively by mankind in its genes. On the other hand, we regard the family and its reproductive activity as a fundamental moral base for our social order. Thus the dilemma. With molecular genetics and advancing technology therein, the problem becomes still more grossly magnified.

Who decides? That is a political question. Whether it is to withhold certain individuals from reproduction, to cause changes in individuals, or to direct by deliberate design certain characteristics in human beings before birth or even before conception, we are talking about a political question that sounds to democratic ears like invasion of privacy and the rights of the individual, to say the least. Is there any individual or institution in the recent history of man possessing the wisdom to direct man's genetic destiny in this way? Yet the question must be dealt with as some aspects of this technology become available.

SUGGESTED READING

1. Phillip Handler, ed., *Biology and the Future of Man*, Oxford University Press, New York, N.Y., 1970. This volume describes the impact of biology on man with great emphasis upon molecular genetics.

2. James D. Watson, *Molecular Biology of the Gene*, W. A. Benjamin, Inc., New York, N.Y., 1965.

3. James Watson, *The Double Helix*, Atheneum Publishers, New York, N.Y., 1968. An account of the formulation of DNA structure by Francis Crick and the author. As well as its scientific importance, the account is highly personal and describes some of the positive and negative aspects of the interactions between scientists.

4. R. P. Levine, *Genetics*, 2nd ed., Holt, Rinehart and Winston, Inc., New York, N.Y., 1968. An account of the fundamentals of genetics.

5. G. E. Moore, *The Cancerous Diseases*, Wadsworth Publishing Company, Inc., Belmont, Calif., 1970. This book is an elementary description of cancer.

6. B. Issekutz, *The Chemotherapy of Cancer*, Akademiai Kiado, Budapest, 1969. An account of the approaches and results of chemotherapy.

7. "Man into Superman: The Promise and Peril of the New Genetics," *Time*, **97**, 33–52, April 19, 1971. This is a detailed but readable account of the developments of molecular biology. It deals with the biological developments and speculates upon their social ramifications.

CASE STUDY 4: THE CHEMISTRY OF CONSCIOUSNESS AND THE INTELLECT

Having looked at the chemistry of how a nerve cell functions (p. 249), we now move to a more difficult question in neurophysiology. How can we describe the phenomenon of intelligence that man exhibits? Man and other animals learn from experience. In learning they somehow store information. It seems possible that some sort of circuit is established among several of the neurons in the brain. Once the circuit is established, it remains intact, though subject to degradation with time. The other aspect of intelligence is the ability to tie these individual pieces of memory together, devising new circuits by this integration. This is what we call *rationalization* or *thought*. We are far from understanding this most important phenomenon, but there are a few hints appearing that suggest how this may be happening. They point toward a tie-in between learning and protein synthesis.

It is possible to teach flatworms a simple task, such as to respond to a light. If these worms are fed to untrained worms, the latter appear to learn faster than the first ones did. Was there some substance in the consumed worms that their brothers could use in their own learning? There are strong suggestions from experiments that amino acids and RNA are involved. Rats who have been trained show changes in their RNA content in the brain when compared with untrained control rats. Moreover, it appears that in one landmark experiment a specific substance has been isolated that we can associate with a specific learning procedure.

Rats are by nature nocturnal, but it is possible to reverse their normal behavior, in effect, to train them to fear the darkness and to seek light. If the brain fluids of rats trained to fear darkness are removed and injected into other rats, the effect is astonishing. The new rats, without training, have the same fear of darkness that the old ones had. The strong inference from the experiment is that a molecule is present that was transferred to the new rats. Chemical analysis of trained rat brain extracts produced a single compound that had the same effect on untrained rats. It is known as *scotophobin*, from Greek words meaning *darkness* and *fear*. It is a simple protein consisting of a sequence of 15 amino acids.

If a molecule is involved in learning in this instance, then it is probable that it is involved for other species. The role of RNA in the process could be as a synthesizer of the protein. How the experience of a sensation is transferred to an RNA molecule and how it or a protein serves to provide a linkup of neurons are questions that have not been answered. To some readers this may indeed smack of "1984", for it appears that molecular conditioning or training of individuals could someday be accomplished by administering the proper chemical. As psychology becomes increasingly a field for molecular study, we shall find this dilemma more and more with us. Advance in scientific knowledge again presents society with a new kind of question.

There are also some hopeful aspects of this molecular psychology. We are approaching understanding the chemistry of disease of the nervous system. In the case of schizophrenia, it has been established that a key enzyme is absent. By restoring the enzyme the disease may be controlled.

SUGGESTED READING

1. W. C. Corning and S. C. Ratner, ed., *Chemistry of Learning—Invertebrate Research*, Plenum Press, New York, N.Y., 1967.

2. William L. Byrne, ed., *Molecular Approaches to Learning and Memory*, Academic Press, New York, N.Y., 1970.

3. Georges Ungar, ed., *Molecular Mechanisms in Memory and Learning*, Plenum Press, New York, N.Y., 1970.

These three volumes present summaries of the first effects dealing with this field.

THE SCIENCE ESTABLISHMENT

Organization and Funding of Research

It is important to consider the substance of science, the ideas and experimental findings that have impact on our lives, and this is what courses in chemistry and other sciences have always attempted. These ideas are the usual substance of chemistry courses, and we have dealt with many of them in this book. In the last section we shall look at the structure of science itself, at how scientists communicate their work and interact with one another. Perhaps of greater importance to both the scientist and the nonscientist is an understanding of how the two interact with one another in the areas of decision-making in science and the interrelation of government in that process.

Before we can proceed further, we need to have a clear understanding about what objectives scientists foresee with respect to their work. Most scientists tend to see research occurring in three broad categories. The first is *basic research.* The objective is to discover new knowledge about nature. The scientist has no practical objective in mind, although he believes that good will ultimately come from the work because man can apply knowledge only after it is acquired. Effective utilization of resources depends on this knowledge, but no one can foresee exactly how before the fact. Scientists are nearly universal in their belief that solving the cancer problem, as an example, requires advances in the basic understanding of the disease. Once that understanding has been gained, it will be far clearer how to proceed with treatment.

The second category is *applied research.* The difference is that there is a clear-cut objective in mind. The objective might be a new product of commercial value, though the research may focus on the objective in a very general way. For instance, the research could involve the synthesis of a new polymer as a potential material for a certain application without a certainty that the substance would actually work. One of the approaches to cancer is chemotherapy. A chemist, in synthesizing compounds of potential chemotherapeutic importance, is engaged in applied research. This example illustrates the gray area between basic and applied research. If a chemist synthesized the same compound with the only objective being to study the synthesis and compound, the research would be in the basic area. The division, in practice, can be only a formalism.

The third area of research, still more practically oriented, is called *development.* In this area the scientist is responsible for bringing to fruition a product, service, or idea for which basic and applied research indicates there is a good chance of success. Much of this work is done by engineers. Again the lines of distribution are not well-defined, nor do they need to be. Most scientists work in all three areas at one time or another in their careers. Developments leading to the atomic bomb serve to illustrate the three areas. Work on the structure of the atom and the nature of radioactivity and nuclear reactions, including the work of Hahn (p. 302), which demonstrated the possibility of a chain reaction, is all basic research. The work of Fermi, in which a chain reaction with release of large quantities of energy was actually accomplished, was applied research with a specific objective in mind. The Manhattan Project, in which a workable nuclear bomb was constructed and tested, was a development program.

Most of the work conducted by private industry is either applied research or development. Since most of the research in an industrial laboratory is directed toward the economic gain of the company, special exceptions to normal scientific operations prevail. For instance, most of the research findings remain the confidential information of the company until patent protection has been assured. The reason, of course, is to prevent competing companies from taking advantage of the research findings. Little of the work is published in the scientific journals.

Research operations in industry are notoriously susceptible to the vicissitudes of the economy. By sacrificing research a company loses in the long run but not over a briefer time. When a recession develops, as in 1970 and 1971, it is likely to affect scientists rather severely and laboratories often "lay off" substantial portions of their staff. Research in industry is extensive in better economic times. It is supported at a level of about 15.5 billion dollars/year.

For that research not supported by industrial corporations, another benefactor must be found, and in the United States and nearly everywhere else that benefactor is government, usually at the national level. The alliance between science and government has always been a little uneasy. Science is uneasy because the power can cause them to prosper or to fail. The politicians tend to hold scientists a little in awe because of the mystique that has arisen around science and its successes and because of problems of communication. The language of science is highly specialized and scientists have not developed great skill in communicating their work to nonscientists.

The result has been a compromise. Politicians in the Congress and the Executive Branch state broad policy, and scientists in advisory positions determine the best approach and what the costs are likely to be. Upon the advice of scientists, funding levels are then determined by Congress. In the coterie of scientific advisors to government, an important place is occupied by the President's Science Advisor, who works along with the other advisors on the White House staff. He has available to him the expertise of other scientific leaders.

Research within government is extensive, and much of it is applied. Generally the objective is one that is perceived to be in the national interest, such as advances in health, food and nutrition, environmental protection, and weapons technology. Many agencies of the federal government are involved in research. A few of the most important of these are listed.

The Department of Defense (DOD)
The National Science Foundation (NSF)
National Institutes of Health (NIH)
The U.S. Department of Agriculture (USDA)
The Food and Drug Administration (FDA)
Atomic Energy Commission (AEC)
National Bureau of Standards (NBS)
Environmental Protection Agency (EPA)
Federal Radiation Council (FRC)
United States Weather Service (USWS)
National Aeronautics and Space Administration (NASA)

Research by these groups is done in laboratories in and near Washington, D.C., and elsewhere. Some of them, particularly NSF, AEC, NIH, and various agencies within the Defense Department, serve as funding agencies for the research conducted in some private laboratories and by universities. AEC in conjunction with many important universities and with some industrial involvement operates National Laboratories at Brookhaven, New York; Argonne, Illinois; Oak Ridge, Tennessee; and Livermore, California.

The aims and research interests of the agencies above should be obvious. One, The National Science Foundation, bears a special relationship to the scientific community, however, because it is the agency charged specifically with the task of sponsoring basic research. It also has a major role to play in science education. The NSF was founded by an act of Congress in 1950. Its annual budget has grown steadily from 15 million to 600 million dollars for fiscal 1972. NSF allocates nearly half of this to basic research principally at academic institutions. About one-quarter of its budget is allocated to science education. The rest is scattered among various specific programs. NSF policy can effect strong influence on activities and vitality of universities. Two decisions recently taken by NSF are of great importance to its clients, the universities. These were to reorient itself somewhat away from basic research by initiating efforts directed toward specific missions. A new arm of NSF, the RANN program (Research Applied to National Needs), instituted in 1971, is aimed at research in several defined areas. These are

1. Environmental Systems and Resources, including study of the environment on a regional basis, problems of trace contaminants, and weather modification studies.
2. Social Systems and Human Resources, primarily focusing on problems of urban centers.
3. Advanced Technology Applications in the fields of energy resources and urban technology. Initial problem areas also stress development of applications of enzymes to research and other advances in industrial technology. In addition to these three broad areas, provision is made for more basic research in the RANN program.

Another major change in NSF in the early 1970's was a decision to change emphasis in the area of graduate education in science. This has been based on the fact that a surplus of Ph.D. recipients, exceeding the demand for both academic positions and industry, had developed. NSF support in the past has been in the form of fellowships that supported a student by paying his tuition costs and his living expenses. By making fewer graduate fellowships available, NSF caused the number of graduate students for whom support is available to decrease. In this fashion, the NSF hopes to bring the number of graduates into line with the number of positions available to them.

The Academic Scientist

The scientist has played his role in higher education in the United States since its beginnings. As both the academic institution and science developed, the number and diversity of scientists on campus has increased. In the typical nineteenth-century college, emphasis was usually on classical learning. By the Civil War an area of natural science as a separate field had emerged in the curriculum, although the specific divisions, chemistry, physics, biology, and so on, were not always distinguished until near 1900. Hence, the typical natural science professor had to be familiar with several science areas. Since then the trend has been toward more and more specialization. The establishment of departments and subdivisions of departments has occurred. Most chemists see themselves as biochemists or analytical, physical, organic, or inorganic chemists. Most are not thoroughly familiar with contemporary research in the other areas, nor do they consider it very important to know about the other areas. They usually find it a struggle to keep up with their own field.

Higher education underwent great change after the Civil War because of the Morrill Act of 1862, which established one land-grant college in each state. The objective of the land-grant college was to encourage agriculture and other applied sciences. Land-grant colleges developed schools of agriculture and engineering. These schools also operated applied research laboratories as a service to farmers in the respective states. This gave impetus to the development of science to supplement the training and research activity in the applied science areas, and science prospered as these universities grew to be among our largest and most distinguished.

Science graduate education began to develop in the early twentieth century and gained impetus after World War I. Before that time scientists had usually obtained their graduate education in Europe. Germany had the most advanced chemical industry, and there was a correspondingly flourishing educational program in chemistry there. Americans and others frequently studied there, and an advanced degree obtained in the United States was considered less prestigious. The pace of growth of graduate education accelerated after World War II in the United States and expanded still further in the 1960's, where 2,000 doctoral degrees awarded in 1 year in chemistry is characteristic.

The accomplishments of graduate study are noted by the awarding of two degrees, the Master of Arts or Science (M.A., M.S.) and the Doctor of Philosophy (Ph.D.). The Master's Degree requires 1 to 2 years beyond 4 years of undergraduate study. The Ph.D. is completed with 3 to 5 years of study beyond the B.S. Degree. The principal difference is the emphasis on research in doctoral study. Early in his doctoral career the student associates himself with a professor in his department whose research is of interest to him.

He undertakes a research problem that they agree upon, and after completion the student submits a thesis on the research. During his career the student is evaluated many times and in many contexts by the faculty with their chief concern being whether the student has demonstrated competence in research. He is also employed and paid a modest stipend by the department as a teaching or research assistant. He is usually not required to pay any tuition or fees.

The Ph.D. Degree has assumed such importance in American education that it is considered the "union card of academia". Most faculty members in colleges and universities possess the degree or are in the final stages of degree work. In principle, all degree holders have been subjected to and have passed the same test; they have demonstrated the ability to carry out a program of independent research, and the faculty believes that they will continue to demonstrate this ability in their career.

Once having completed the doctorate, a chemist must choose an industrial, government, or an academic career. About 60% choose industry. The academic followers may elect to receive further training at the postdoctoral level by working in association with a professor at another institution. After 1 or 2 years as a "post-doc," he will move into an academic position as a junior-level professor. At this point several possibilities for quite different careers await him. The choice surrounds whether he wishes to work in an institution that is primarily undergraduate in its emphasis, which means that lecturing and classroom activity will occupy the major part of his time, or to associate himself with a college with a graduate program, in particular, with a Ph.D. program. If the chemist chooses this, classroom and undergraduate teaching will probably be subordinate to his research, in particular to working with and evaluating graduate students, to publishing research results in chemical journals, and to securing support from NSF and other agencies for his work.

In the small college or university, or in the junior college, teaching "loads" are high. Typically, a professor is responsible for teaching three courses plus the accompanying laboratory, and his time is taken up in preparing and delivering lectures, grading examinations, and preparing chemicals and equipment for the laboratory sessions. In addition, he will serve on departmental and university committees.

In the large university the professor will teach only one or two courses. If his teaching is at the undergraduate level, he will be assisted by graduate students in the laboratory and by stockroom supervisors who make sure that the laboratories are properly supplied. His grading work is usually done by graduate students as well. A major portion of his teaching activity is in the informal day-to-day contact with graduate students. The relationship is not unlike the master–apprentice arrangement in that the professor seeks to instill his own knowledge, experience, and philosophy into the student as the two work together on a common problem. The professor in a large school must also deal with administrative work and with responsibility for financial support of his research and his graduate students.

The evaluation of a professor's work is done by the other members of his department, principally. After a probationary period of 3 to 6 years, a decision is made on the question of faculty tenure. Under the tenure system a faculty member has a continuing appointment, meaning that he will not be dismissed unless he commits a serious crime or indiscretion, and barring severe financial difficulty for the college. The idea of tenure is to enable the teacher to function as he sees fit in his teaching and research without fearing for his job, on the grounds that he will be more effective with this freedom to pursue his ideas and research wherever they lead him. Evaluation for tenure and also for promotion and salary increases will depend on how effectively the faculty member is judged to be carrying out his responsibility. In the small school this means that he has demonstrated effective teaching ability and has done his share of department and university administrative work. In the large university the evaluation is based primarily on research productivity and quality, and this is judged on the basis of publication of his research in the scientific literature, his success in securing research grants, and his effectiveness in directing graduate students in research. Upon these evaluations, a judgement concerning his potential for continued research effectiveness is made.

In both publication and grant proposals, the critical point is the review of the work by a chemist's peers. When a research problem has been brought to fruition, a paper is written describing the experiments and interpreting their meaning. This is sent to an editor of one of many chemical journals. If the work is in a field of chemistry that the journal covers, the editor will send the manuscript to one or more referees who are experienced in the area or discipline that the paper deals with. The referees comment on the manuscript, evaluate its importance and correctness, and advise the editor whether or not to publish the manuscript in his journal. The editor will either accept or reject the work, or he may recommend revision to the author.

With research proposals the chemist will again face referees, in this case panels convened by the granting agency. In his proposal the chemist elaborates upon his research idea. He outlines his experimental plans and how the results will have importance. He also submits a budget in which provision is made typically for needed equipment, for personnel including some support of graduate students, for overhead to the university, and for the costs of publication. The decision of the granting agency will be based upon the review of the panel's evaluation and the availability of funds.

In the mid-1960's the system that we have just described seemed to be working well. College enrollments were burgeoning and, as a consequence, there was a brisk demand for faculty, which in turn encouraged increased training of graduate students. Money for purchase of equipment, graduate fellowships, and research was readily available, and thus a period of rapid growth had been experienced. The future of chemical academe seemed well assured.

The experience of the 1970's has abruptly dispelled this euphoria. In the first place, the economic recession of 1970 and 1971 took a heavy toll in the chemical industry. Few new jobs developed there and many working industrial chemists found themselves unemployed. Problems developed too at the colleges and universities. Private schools found that inflation had created a cost-price squeeze such that they were having difficulty paying fair salaries to faculty and at the same time keeping tuition in bounds. The solution was usually to hold back on hiring new faculty. In state universities the problem was a reluctance of the state legislature to increase the levels of funding. In many instances these schools were forced to adopt freezes on new growth of the same sort that private colleges had done. Also, the change in philosophy at NSF meant a de-emphasis on graduate education and a reorientation in university and departmental research goals may become required.

Criticism of chemistry and other science departments also came from students. The criticism at large universities was that the undergraduate teaching program had become too much the poor stepchild of the department, with graduate education receiving all the attention. Some students demanded more effort be given to undergraduate training and others even expressed the desire for student involvement in such areas as curriculum evaluation and faculty promotion and tenure decisions. Reforms such as changing grading and examination systems were often advocated. Most institutions have undertaken studies of these criticisms and varying changes have occurred. Their effectiveness will be clearer with time.

The other criticism that came forth in the early 1970's was the charge of irrelevance. In both the teaching and research of chemistry, departments tended to view their discipline in terms that neglected the interaction of science with the total society, in particular in terms of technological innovation. It was pointed out that in too many instances scientific innovation had done great harm. The effect of the automobile on the city, pesticides, and the development of industrial concentrations and consequent water pollution were all examples of this. The adverse effects of the development of military technology was severely criticized.

Scientists were startled and chagrined by these charges, which they felt went against the most fundamental tenets of established science. Every chemist believes that fundamental research is the basis on which technological advance depends, and he also believes that technological advance is good. The question of how knowledge is to be applied in technology was of concern, but scientists left that question to others on the grounds that the decision was economic or political. There is a flaw in this argument that illuminates a dilemma for our culture and our time. While the scientists vacate the arena of decision to others, the decision-maker badly needs enlightened help from scientists to fulfill his role. It is hard for him to understand the complexities of scientific reasoning and the implications of new knowledge on human life. Members of Congress, for example, have great difficulty understanding scientific ideas. Even the language that the scientist uses may present a barrier to comprehending.

Clearly better understanding between scientists and nonscientists is needed as we move to clean our environment and to evaluate better the role of man in his environment and in his varying societies. Both sides need to learn to express themselves in a straightforward way to laymen and to consider potential effects of their work more carefully. On the other side, more scientific literacy is needed on the part of politicians, journalists, and the general public. Given these two developments, effective and careful methods of taming technology can be devised to the benefit of man and society.

The chemistry profession is responding to the new situation. More concern for applied research in academic laboratories has been expressed. Significant changes in curriculum have been devised, especially in the area of chemistry courses for nonscientists. Development of new degree programs that emphasize teaching more and training chemists to work in government and local education are being considered. A chemist trained in this way will have a broader background than the pure research orientation of the Ph.D. recipient, and he should be able to make a different kind of contribution to society.

On the industrial side it is clear that decisions in the future will involve great consideration of the environment. Industries face a problem of being caught in the middle by competition pressures and production costs and by consumer demands. They will tend to produce products giving these factors higher priority than environmental factors or other social concerns. The marketplace is amoral in capitalism. It is possible to change demand patterns by education or by law, however, and we shall see this in the future. Industrial research will be central in solving problems related to this new orientation.

SUGGESTED READING

1. Lewis M. Branscomb, "Taming Technology," *Science*, **171**, 972 (1971). This paper discusses the problem of technological control in terms of the interface between scientist and nonscientist. The author is director of the National Bureau of Standards.

2. Allan M. Cartter, "Scientific Manpower for 1970–1985," *Science*, **172**, 132–140 (1971). This paper describes recent changes in supply-demand patterns for Ph.D. scientists and projects those changes into the future.

3. V. R. Potter, *Bioethics: Bridge to the Future*, Prentice-Hall, Inc., Englewood Cliffs, N.J., 1971. This book describes the problem of the interface of science and ethics, in particular as it relates to biology.

4. R. B. Fischer, *Science, Man and Society*, W. B. Saunders, Co., Philadelphia, Pa., 1971. A discussion of the role of scientists in society.

5. Garrin McCain and E. M. Segal, *The Game of Science*, Brooks/Cole Publishing Co., Belmont, Calif., 1970. A whimsical look at the machinery of contemporary science.

6. D. S. Greenburg, *The Politics of Pure Science*, The New American Library, Inc., New York, N.Y., 1967. This book was written by a former political editor of *Science*, the weekly journal of the American Association for the Advancement of Science. It details the post-World War II development of the science establishment.

Postscript

I once had a cow that jumped over the moon,
Not on to the moon but over.
I don't know what made her so lunar a loon;
All she'd been having was clover.

That was back in the days of my godmother Goose.
But though we are goosier now,
And all tanked up with mineral juice,
We haven't caught up with my cow.

From the poem
"Lines Written in Dejection
on the Eve of Great Success"
by Robert Frost, 1962.

from The Poetry of Robert Frost,
edited by Edward Connery Lathem,
Holt, Rinehart and Winston, Inc., 1969,
Jonathan Cape Ltd.
and the Estate of Robert Frost

Glossary

Each of the terms included in the glossary is identified by an asterisk at its first occurrence in the text. The purpose of the glossary is to define more fully some concepts and words which are used in the book. You may wish to refer to a standard general chemistry book for further discussion.

Alcohol   A class of organic compound in which an hydroxyl group ($-\ddot{\text{O}}-\text{H}$) is attached to a hydrocarbon group; for example, ethyl alcohol $CH_3-CH_2-\ddot{\text{O}}-H$. However, phenols, carboxylic acids, and a few other classes are not considered to be alcohols even though they contain hydroxyl groups.

Alpha (α) particle   A particle emitted from a nucleus during some forms of radioactive decay. The particle has a mass of 4 amu, and a charge of $+2$; thus, it is equivalent to a helium nucleus.

American Association for the Advancement of Science (AAAS)   A national organization of scientists from many disciplines in the physical, biological, and social sciences. Its principal function is to publish and otherwise disseminate scientific advances. It also encourages support of scientific research and publishes the weekly magazine, *Science*.

American Chemical Society (ACS)   The principal organization of chemists and chemical engineers in the United States. Its principal objectives are the dissemination of research advances in chemistry, the encouragement of useful applications of chemistry, chemical education at all levels, and the professional and economic advancement of chemists.

Amide A class of organic compounds characterized by the presence of the partial structure:

A simple example is the compound N-methylacetamide

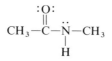

Amines Derivatives of ammonia in which hydrogens are substituted with organic groups. Depending upon the number of hydrogens substituted, the amines are known as primary, secondary, or tertiary amines.

$$:NH_3, \qquad CH_3\overset{..}{N}H_2, \qquad (CH_3)_2\overset{..}{N}H, \qquad (CH_3)_3N:$$

ammonia	methylamine (primary)	dimethylamine (secondary)	trimethylamine (tertiary)

Amino acids A class of organic compounds containing both amine and carboxyl groups. Amino acids of the structure

$$R-\underset{\underset{NH_2}{|}}{CH}-CO_2H$$

are the monomeric units which form proteins. This is accomplished by forming an amide or peptide bond between the amine group of one unit with the carboxyl group of another. Different amino acids have different groups attached at the R position.

Ångstrom A unit of length, 10^{-10} meters, abbreviated Å. Such a unit is convenient to describe distances in atoms and molecules. For example, the distance between hydrogen nuclei in H_2 is 0.75 Ångstroms. The carbon–carbon single bond distance in most hydrocarbons is about 1.54 Å.

Antibodies Substances which are synthesized by the body to act against invading disease organisms. By the growth of antibodies, immunity is developed.

Antibonding orbital A molecular orbital of high energy which, when occupied by electrons, acts to decrease bonding between nuclei. Although antibonding orbitals are usually not occupied, an electron is sometimes promoted to one by energy of the appropriate frequency, leading to molecular decomposition or other reactions.

Atomic mass units (amu) The scale of mass most frequently used in describing atomic weights. One atom of the ^{12}C isotope is taken to have a mass of exactly 12 amu, and the weight of the atoms is established by comparison. One atomic mass unit is equivalent to 1.66×10^{-24} g. A synonymous term is dalton.

Biochemical Oxygen Demand The weight of dissolved oxygen consumed in the degradation of organic matter present in natural waters.

Biogenesis The synthesis of naturally occurring compounds in living tissues. Starting from simple molecules, a series of reactions involving enzymes as catalysts is usually required.

Biomass The total matter constituted in all living things, both plant and animal, in a given system; for example, in a lake or a forest, or even the entire earth.

Bond vibrations In molecules there is constant motion of the atomic nuclei. Vibrational motions in which the nuclei move back and forth along the internuclear axis as well as twisting and bending vibrations are of interest. Vibrational frequency may change upon absorption of energy in the infrared region of the spectrum.

Carboxyl group The functional group present in carboxylic acids. Its structural representation is

e.g., in benzoic acid

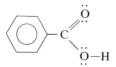

Chelating agents A ligand coordinating with a central ion through two or more atoms. (*See also* **Ligands.**)

Coenzyme Substances which function in conjunction with an enzyme in a biochemical reaction. (*See also* **Enzyme.**)

Complex The product of association of a Lewis acid and Lewis base, particularly when the Lewis acid is a metal cation. Among many examples cited in the text is $Ag(NH_3)_2^+$.

Compost Organic matter, such as food waste, plant matter, or manure, which is decomposed by bacterial action in the absence of oxygen to form a humus soily material, of value as a fertilizer.

Copolymer A polymer which contains two or more monomeric units linked together either in an alternating or a random fashion.

Covalent bond A bond formed by the sharing of an electron pair between two atoms.

Critical mass In nuclear reactions caused by neutron bombardment, the mass of reacting material which is large enough to produce sufficient neutrons to sustain the reaction by neutrons released from fissioning nuclei continuing the chain reaction.

Cupric ion Copper in its $+2$ oxidation state, Cu^{+2}, as in CuO, cupric oxide.

Cuprous ion Copper in its $+1$ oxidation state, Cu^{+1}, as in Cu_2O, cuprous oxide.

Cyclotron A device for separating charged particles of different masses by accelerating them through a magnetic field in which they are deflected in differing degrees as a function of their mass. This technique was one used with partial success in separating U^{235} from natural uranium.

Distillation A process for purification or separation of a mixture in which some components are vaporized and then recondensed into another vessel. In this fashion more volatile components may be separated from less volatile ones. (*See also* **Fractional distillation, Volatility**.)

Electrolysis A chemical reaction caused by the passage of electric current through a solution or other reacting medium.

Electromagnetic radiation The propagation of energy as waves in the form of oscillating electric and magnetic fields.

Electron A basic atomic particle of negative charge balancing the positive charge of the proton. They exist in the peripheral areas of the atom and are much lighter in mass (9.11×10^{-28} g) than the proton.

Electronic transitions The excitation of an atom or molecule by the promotion of an electron from one energy level to another higher one. Most such transitions take place at energies in the visible and ultraviolet spectral regions.

Endothermic reaction A chemical reaction in which there is a net absorption of kinetic energy, converted to the potential energy of new chemical bonds. (*See also* **Exothermic reaction**.)

Energy levels (*See* **Orbital**.)

Enzyme Proteins which act as catalysts for chemical reactions that occur in living things. Enzymes are usually quite large molecules constructed of a sequence of amino acids. They enable biochemical reactions to occur at far faster rates than they could otherwise. (*See also* p. 317 in the text.)

Epoxide A class of organic compounds in which is present an oxygen-containing, three-membered ring; e.g.,

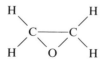

known as ethylene oxide, an important industrial monomer. By substitution with other groups for hydrogen, many different compounds are possible.

Equilibrium In chemistry a system in which two reverse reactions are occurring such that there is no net change in the amounts of any of the reactants or products present.

Equilibrium constant A chemical reaction at equilibrium is usually described in terms of the relative quantities of reactants and products by the equilibrium constant. For a reaction

$$aA + bB \rightleftharpoons cC + dD$$

where A, B, C, and D are chemical components and a, b, c, and d are coefficients in the equation, the equilibrium constant K in terms of the concentrations of the components is:

$$K = \frac{[D]^d[C]^c}{[A]^a[B]^b}$$

each component concentration raised to the power of the coefficient. The products appear in the numerator, the reactants in the denominator.

Ester   A class of organic compounds characterized by the presence of the partial structure

A simple example is the compound ethyl benzoate

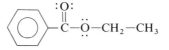

Excited state   The condition existing in an atom or molecule in which energy is absorbed and an electron is promoted from the lowest energy level to an unoccupied higher energy level. When all electrons are in the lowest energy levels, the atom or molecule is said to be in its *ground state*.

Exothermic reaction   A chemical reaction in which potential energy is released as heat to the surroundings. (*See also* **Endothermic reaction, Potential energy**.) The metabolism of glucose is an exothermic process.

Ferric ion   Iron in its $+3$ oxidation state, Fe^{+3}, as in ferric oxide Fe_2O_3.

Ferrous ion   Iron in its $+2$ oxidation state, Fe^{+2}, as in ferrous oxide FeO.

Fractional distillation   A more complicated distillation process in which the distilled materials are separated into fractions on the basis of boiling temperature ranges. The process is used to separate out various fractions in petroleum. (*See also* **Distillation**.)

Free radical   A molecule which possesses one or more unpaired electrons. Most free radicals are unstable, since there is a strong impetus for electrons to exist in paired states. However a few free radicals, such as NO and NO_2, are stable at room temperature, and many are stable at lower temperatures.

Frequency   Used to describe electromagnetic or wave radiation in terms of the number of waves which pass a given point in a given time. The value is given in cycles/sec or hertz.

Glycosidic linkage   The combination of two units of glucose or other carbohydrates with the loss of a molecule of water. For example, maltose forms by the glycosidic linkage of two molecules of glucose. Many such linkages produce polymeric carbohydrates such as starch and cellulose constituents.

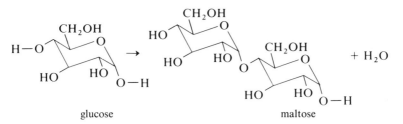

Hertz Unit of frequency; cycles per sec. For radiation in the radiofrequency region, the term megahertz (millions of hertz) is used.

Hydrogen bonding A weak bond formed between a hydrogen on one molecule with an electronegative element on another. Hydrogen bonding occurs with hydrogens bonded to the more electronegative elements, F, O, S, Cl, such that the hydrogen involved bears a partial positive charge.

Infrared Frequencies of electromagnetic radiation lower in energy than that of visible light. Much of the heat radiated into space from the atmosphere is in the infrared region. Infrared spectroscopy is an important tool for the study of molecular structure (*see* chapter 7).

in vivo A biochemical term describing the results of a chemical reaction or other process in living tissue. A second term *in vitro* (in the glass) refers to similar reactions which are carried out in the laboratory under conditions which attempt to match the living cell.

Ion An atom or molecule which bears an electric charge, either positive or negative.

Kilocalorie A unit of heat frequently used in chemical reaction; 1000 calories, a calorie being the quantity of heat required to raise the temperature of one gram of water from 14.5°C to 15.5°C. A kilocalorie is of the same magnitude as the nutritional Calorie, which is always capitalized to distinguish it from the smaller unit.

Kilowatt A unit of power, or energy per unit time. One kilowatt is equivalent to an energy output of 0.239 kilocalories/second. A *kilowatt-hour* is an energy unit, the total energy produced per hour at a rate of one kilowatt, equivalent to approximately 859 kilocalories.

Lattice The arrangement of atoms, molecules, or ions in the solid state. In crystals the lattice possesses a very orderly arrangement.

Lewis acid Substances which are electron-deficient, usually capable of coordinating with Lewis bases by accepting electron pairs.

Lewis base Substances which are electron-rich, usually capable of donating an electron pair to a Lewis acid.

Lewis structure A representation of a molecular structure in which all valence electrons are represented by dots in writing the formula. For methyl chloride (CH_3Cl) the Lewis structure is

$$
\begin{array}{c}
H \\
H : C : \overset{..}{\underset{..}{Cl}} : \\
H
\end{array}
$$

The method is named for G. N. Lewis, a distinguished American chemist, who formulated the idea of an octet of valence electrons for covalent structure of the second-row elements.

Ligands Electron-rich substances which can coordinate with electron-deficient species, such as metal cations, by donation of an electron pair. (*See also* **Lewis base**.)

Limnology The study of ponds and lakes and changes they undergo due to natural and man-induced processes.

Lipid   Substances occurring in plant and animal tissues which are soluble in organic solvents, such as benzene or ether. In this category are fats, steroid hormones, and essential oils of plants.

Logarithm   An exponent or power to which a given base is multiplied to produce a given value. A simple example is: $100 = 10^2$, where 2 is the logarithm required to produce 100 given a base of ten. Many scientific relations are based upon a natural logarithmic scale in which the base is e having the value 2.718, rather than base ten. The natural logarithm is derived from calculus.

Meta   Benzene compounds in which two substituents are attached to the benzene ring at two carbon atoms separated by a third carbon atom; abbreviated as m, e.g., m-dichlorobenzene. (*See also* **Ortho, Para**.)

Metal   Those substances appearing in the upper left portion of the periodic table. As elements they tend to be lustrous, malleable solids conducting heat and electricity. They also tend to be oxidized readily by loss of electrons.

Microns   A unit of distance, abbreviated μ, to describe wavelengths in the infrared region of the spectrum. A micron is equivalent to 10^{-6} meters.

Mycelle   An area within the body of a solution in which a solute with surfactant properties forms an agglomerate structure.

Natural logarithm   (*See* **Logarithm**.)

Nomenclature   A subdiscipline of chemistry whose objective it is to determine rules and practices to enable the unambiguous naming of pure chemical substances.

Nonbonding valence electrons   Electrons present in the valence octet of an atom, but which are not shared with other atoms. In water, for example, there are four nonbonding electrons on the oxygen.

Olefin   An unsaturated compound (containing carbon–carbon double bonds). An example is propylene

Orbital   The region of space around an atomic nucleus or nuclei where there is a high probability of finding an electron; also, the region of high-electron cloud density. The term is used interchangeably with energy level and is used with reference to both atoms and molecules.

Organic Synthesis The activity of chemistry in which naturally-occurring or synthetic organic compounds are prepared in the laboratory.

Ortho Benzene compounds, in which two substituents are attached to the benzene ring at adjacent carbon atoms. The substituents may be alike or different. Abbreviation used is *o*, e.g., o-dichlorobenzene. (*See also* **Meta, Para.**)

Oxidation A term which has different meanings in different contexts; the loss of electrons making an atom more electrically positive, or the increase in valence number or oxidation state; the opposite of reduction.

Oxidation potential The energy required to remove an electron from an atom or molecule. On p. 164, reference is made to the fact that oxidation potentials differ for converting Fe, Cu, Ag, and Au, to their respective ions, this being used as a basis for separation during electrolysis.

Oxidation state For a simple ion, the electric charge. Thus, Cl^- has an oxidation state of -1; Fe^{+2}, $+2$. For complex ions and organic molecules, the oxidation state is determined by assigning both electrons in a covalent bond to the more electronegative element and then determining the charges. For $SO_4^=$, each oxygen is -2 and sulfur is $+6$. For methane, CH_4, the hydrogen is assigned an oxidation state of $+1$, the carbon is then -4.

Para Benzene compounds in which two substituents are attached to the benzene ring at two carbons across the ring from each other. Abbreviated as p; e.g., p-dichloro-benzene. (*See also* **Ortho, Meta.**)

Parts per million (ppm) A unit of concentration used for components present in small quantities. The concentration may be based upon volume, weight, or molar quantities of the component in the total sample. In dealing with air pollution, volume concentrations are usually cited. For lesser concentrations parts per billion (ppb), or even parts per trillion (ppt), are frequently used.

pH A number relating the degree of acidity and basicity of an aqueous solution. Most solutions lie in a range of pH from 0 to 14. Lower pH reflects high acid (H_3O^+) concentrations. Higher pH values are highly basic (^-OH) solutions. A pH of 7 reflects neutrality, in which the concentration of H_3O^+ and ^-OH ions are equal.

The mathematical definition relates to the concentration of H_3O^+ on a negative logarithmic scale.

$$pH = -\log[H_3O^+]$$

$$\text{For } [H_3O^+] = 10^{-5}, pH = 5$$

Phenol   A class of organic compounds in which an hydroxyl group (^-OH) is joined to an aromatic ring (e.g., benzene). The simplest such compound, $\langle\!\bigcirc\!\rangle\!-\!\ddot{O}H$, is also known as phenol. (*See also* **Hydroxyl group**.)

Photosynthesis   The formation of glucose and other carbohydrates by reaction of carbon dioxide and water catalyzed by chlorophyll in the presence of sunlight. The chemical details of the process are just beginning to be understood. The overall chemical equation for photosynthesis is shown on p. 19.

Polar Bond   A covalent bond in which the electron pair is unevenly distributed with respect to the two nuclear centers. Thus, small positive and negative charge centers develop within the molecule. (*See also* **Covalent bond**.)

Polyolefin   A polymer formed whose monomeric unit is an olefin.

Potential energy   Energy of position or condition. Potential energy is related to chemical bonding because the atomic positions are relatively fixed with respect to one another. When chemical bonds are broken and formed in a chemical reaction, there is a net change in potential energy. Some may be converted to kinetic energy, released as heat, or kinetic energy may be absorbed from the surroundings and stored as potential energy in new bonds.

Protein   Natural polymers consisting of amino acids condensed together to form long units. (*See also* **Amino acids, Enzymes**.)

Protein structure   Protein structure is thought of in four levels. *Primary structure* is the simple amino acid sequence. *Secondary structure* is the twisting of the amino acid chain into a coiled structure known as the α-helix. *Tertiary structure* is the shaping of the α-helix by interactions among various parts of the helix. *Quaternary structure* is the combination of two or more subunits into the total protein structure.

Proton   A basic particle of nuclear structure bearing a positive charge. The hydrogen nucleus consists of a single proton. Other nuclei consist of a number of protons and neutrons. The mass of the proton is 1.67×10^{-24} g.

Proton-transfer reaction   In general terms, $HA + B \rightleftharpoons HB^+ + A^-$ where HA, an acid, transfers a proton to B, a base, producing the products at the right of the arrow. Specific examples are shown on page 90 and elsewhere throughout the text.

Quantum mechanics   That branch of physics dealing with the behavior of small particles in which their motion is treated mathematically in terms of wave properties.

Reduction   The opposite of oxidation; the addition of electrons or the decrease in oxidation state.

Resolution   The separation of mirror-image optical isomers. Since their physical properties are the same, it is necessary to introduce another asymmetric carbon center by chemical reaction. By reacting mirror image forms R and S with an optically-active compound R', two new forms R–R' + S–R' are produced. The new pair are not mirror-image forms (since the opposite of R–R' is S–S') and they can be separated by chemical methods.

Singlet state   Referring to a molecule in which two electrons possess opposite spins orientations. (*See also* **Triplet state**.)

Soft and hard acids and bases   A theory extending the concept of Lewis acids and bases. Soft bases and acids are characterized by large size and easy polarization of their electron clouds in contrast to hard acids and bases. The theory holds that soft acids and bases have a stronger affinity for one another than for a harder counterpart. (*See also* **Lewis acid**, **Lewis base**.)

Solubility   The maximum quantity of solute which may be introduced into a given quantity of solvent while yet maintaining a homogeneous solution. Solubility is reported as a numerical quantity; for example, moles of solute per liter of solution, grams of solute per kilogram of solvent, grams per liter, ppm, etc.

Surfactant   A substance that changes the surface characteristics of a solvent, usually by lowering surface tension. Both soaps and synthetic detergents act as surfactants.

Technology   The systematic application of knowledge to generate new materials, machines, or other developments.

Tetravalent   An atom forming four covalent bonds to other atoms. In almost all stable organic compounds, the carbon atoms are tetravalent.

Triplet state   Referring to a molecule in which two electrons possess parallel spin orientations. (*See also* **Singlet state**.)

Ultraviolet (UV)   Radiation of shorter wavelengths, hence higher energy, than visible light. The near ultraviolet range occurs at wavelengths below 4000 Ångstroms to about 150 Ångstroms. Energy of still shorter wavelength lies in the far ultraviolet region. The UV region is useful to the chemist because electronic transitions in atoms and molecules occur at these frequencies and may be used for analytical purposes. (*See also* **Electronic transitions**, **Visible light.**)

Unsaturation   With reference to organic compounds, the term indicates that there is at least one double or triple bond between carbon atoms in the molecule. An unsaturated compound is converted to a saturated one by the addition of hydrogen. (As an example, see p. 56.)

Visible light   Radiation in the wavelength region between 4,000 and 7,000 Ångstroms to which the human eye is sensitive. The violet region occurs at the shorter wavelength end of the region, extending through the rainbow into the red region which lies at the lowest energy portion of the region.

Volatility   The tendency for a liquid substance to be converted to a vapor. A volatile substance will have a high vapor pressure and low boiling point.

Wavelength   In describing radiation, the distance of the wave which is descriptive of the entire repeating wave structure.

Index

AAAS, 9, 335
Abortion, 222, 235
Absorption spectroscopy, 202
Acetaldehyde, 179, 277
Acetamide, 264
Acetate ion, 171, 277
Acetic acid, 7, 83, 153
Acetic anhydride, 110
Acetyl chloride, 292
Acetyl coenzyme A, 277
Acetyl radical, 56
Acetylation, morphine, 292
Acetylcholine, 254–6, 307
Acetylcholinesterase, 144, 256
Acetylene, 179
Acids, in pickling steel, 162
Acidulants, 238
Acrylonitrile, 107
ACS, 9, 335
Addiction:
    alcohol, 274, 278
    barbiturates, 266
    heroin, 290
    marijuana, 287
Adenine, 311
Adenosine diphosphate (see ADP)
Adenosine triphosphate (see ADP)
ADP, 22, 68
Adrenaline, 255-260
Aging, 318
Africa, population, 221
Air pollution:
    analysis of, 205
    chemical industry, 115
    food industry, 124
    incinerators, 113
    Kraft process, 120
    steel industry, 162
Alanine, 243
Alcohol (see also Ethyl alcohol), 273
Alcoholics, 274
Alcoholism, 278
Alcohols, 8, 83, 234, 335
Aldehydes, 54, 206, 256, 279
Aldrin, 136
    analysis, 198
Alfalfa aphid, 151
Algae bloom, 70
Alkaloids, 66, 149, 267, 290
Alkanes, 27
Alloys, 161
Almond moth, 147
Alpha particle, 39, 335

Aluminum, 166
    resources, 168
Aluminum chloride, 112
Amalgam, 177
American Association for the Advancement
    of Science (see AAAS)
American Chemical Society (see ACS)
American Scientist, 9
Amide, 107, 144, 336
Amines, 124, 246, 258, 290, 311, 336
    and alkaloids, 268
Amino acids, 66, 171, 336
    and learning, 322
Amino acid sequence, and RNA, 316
Amino group, 307
Δ-aminolevulinic acid, 173
Ammonia, 65
    in fertilizer, 76
    as Lewis acid, 171
Ammonium ion, 255
Amobarbital, 262
Amphetamine, 258, 297
    effects of, 261
Amylose, 21
Amytal, 262
Anaerobic bacteria, 70, 187
Analgesic effects, of morphine, 293
Analysis, of pesticides, 195
Analytical chemistry, 194
Androgenic hormones, 224
Angstrom, 336
Animal wastes, 74
Anode, 164-166
Anopheles mosquito, 130, 150
Anthrax, 308
Antiacetylcholinesterase, 307
Antibodies, 319, 336
Anti-bonding orbital, 54, 336
Antifreeze, 102
Anti-knock properties, 50
Antioxidants, 109, 239
Aphids, 151
Apholate, 146
Appetite depression, amphetamines, 261
Applied research, 324
Army Corps of Engineers, 211
Aromatic compounds, 51, 102, 255
    279, 286
    in lignin, 122
Artificial:
    food colors, 240
    sweeteners, 241
Asia, population, 221

Aspartic acid, 243
Asphalt, 100
Aspirin, 110
Asymmetric carbon center, 258
Athletes, and amphetamines, 261
Atlanta, Georgia, 189
Atmospheric absorption:
  of solar energy, 17
Atomic absorption:
  particulate analysis, 206
  spectroscopy, 204
Atomic mass units, 336
Atomic Energy Commission, 46, 212, 325
ATP, 22, 68
  and nerve cell, 252
Auto exhaust emissions, 59
  analysis of, 206
  standards, 60
Automobile, 47
Autonomic nervous system, 250
Axon, 251

Baby food, and MSG, 244
Bacteria:
  as insect predators, 145
  in sewage treatment, 94
Bacterial agents, 308
Bag house, 162
Bakelite, 108
Barbiturates, 262
  effects of, 266
Basic oxygen process, 161
Basic research, 324
Bauxite, 166
Beef cattle, 74, 145
Beer, packaging of, 193
Benefication, of copper ore, 163
Benzaldehyde, 239
Benzedrine, 258
Benzene, 7, 103, 111–112
Benzene hexachlorides, 135
Benzenesulfonic acid, 86
Beryllium, toxicity of, 183
BHA, 239
BHT, 239
Biochemical oxygen demand (*see* BOD)
Biogenesis, 281, 337
  of Δ -THC, 286
Biogenic amines, 253, 307
  and alkaloids, 269
  effects of morphine on, 293
  synthesis, 257
Biological activity, of optical isomers, 259
Biomass, 39, 74, 337
Biphenyl, 113
Birds, and DDT, 138
Birds, and thallium poisoning, 184
Birth defects, 92
  and herbicides, 154
  and LSD, 272
  and marijuana, 287
Birth rate, United States, 236
Blast furnace, 160
Blood, 122
Blood levels of alcohol, 276
Blue-green algae, 66
BOD, 98, 115, 162, 338
  in food process industries, 122
  in papermaking, 121
Boll weevil, 146
Bonding orbital, 54
Bond vibrations, 337
Bordeaux mixture, 151
Bottled gas, 100

Botulism, 246
Breeder reactors, 44
Builders in detergents, 90
Bureau of Mines, U.S., 191
Butadiene, 102, 108
Butylatedhydroxylanisole (*see* BHA)
Butylatedhydroxytoluene (*see* BHT)

Cadmium ion, 92
  toxicity of, 184
Caffeine, 268
Calcium carbonate, in egg sheels, 139
Calcium dihydrogen phosphate, 239
Calcium hydroxide, 94
Calcium ion, 85
  in hard water, 90
  as nutrient, 169
Calcium phosphate, 94
Cancer, 309, 318
  and cigarette smoking, 319
  and DES, 245
  and LSD, 272
  and marijuana, 287
  and nitrosamines, 246
  research, 319
Cannabidiols, 282-283
Cannabis sativa, 280
Capacitation, of sperm, 230
Caprolactam, 108
Carbamates, 152
Carbaryl, 144
Carbohydrates, 20, 68, 259
Carbon, 120
  in pig iron, 161
Carbonates, 93, 152
Carbon cycle, 26
Carbon dioxide, 25
  atmospheric content, 36
  energy absorption, 37
  fermentation processes, 276
  leavening process, 239
Carbonic anhydrase, 139
Carbon monoxide, 53, 159
Carbonyl group, 262
Carboxylate ion, 85, 171
Carboxyl group, 8, 287, 337
Carboxylic acids, 83, 86, 234, 241, 252
Carboxymethylcellulose, 238
Carboxypeptidase A, 173
Carcinogenic agents, 319
Caspian Sea, 72
Catechol, 255
Cathode, 164-166
Cattle feed, and DES, 245
CBW, 306
Cellulose, 18, 21, 116
Central nervous system, 249
  and morphine, 293
Cervical mucus, and contraception, 231
Chain reaction, 43
Chalcocite, 163
Chalcopyrite, 163
Charcoal, 120
Chelating agents, 90, 337
Chemical & Engineering News, 9
Chemical products, 98
Chloral, 112, 129
Chlor-alkali process, 178
Chlorinated hydrocarbons in biosphere, 136
Chlorine, 118, 306
Chlorobenzene, 112, 129
Chloroethylene, 107
Chlorophyll, 18, 172
Chloroplast, 18

Chloroprene, 108
Chromatography, 197, 281
Chromosomes, 310
Cigarette smoking, and cancer, 319
Cinnabar, 178
Cis-geometry, 282
Cis-polyisoprene, 108
Citric acid, 238
Clean Air Act, 212
Cloning, 321
CN, 308
Coal, 99
    and cadmium, 184
    gasification, 35
    resources, 30
    and toxic metals, 183
Cocaine, 268
Cockroaches, sex attractants, 149
Codeine, 292-293
Codon, 316
Coenzyme, 180, 337
Coke, 159
Coke ovens, 161
Colors, as food additives, 240
Column chromatography, 197
Combination pill, 228
Combustion, 47
    of alcohol, 276
    of fossil fuels, 30
Common ion effect, 120
Communication, and science, 302
Complexes, 337
Composition, of solid waste, 187
Compost, 337
Composting, of solid waste, 188
Compression ratio in auto engines, 52
Compton, Arthur, 304
Computers, in auto exhaust analysis, 208
Concentration gradient, in neuron, 252
Concentration of ions, in neuron, 251
Conductivity, of copper, 163
    of metals, 157
Coniine, 268
Consciousness, chemistry of, 322
Constipation, and heroin, 293
Consumers, 69
Contraception, 222
Contraceptive methods, 221
Contraceptives, long term, 233
Copolymers, 108, 337
Copper, 163
    in IUD's, 232
    resources, 168
Corpus luteum, 225
Cost, of heroin addiction, 290
    of solid waste disposal, 191
Coumarin, 153
Covalent bonds, 81, 337
Cracking of petroleum, 101
Crime, and heroin, 290
Critical mass, 43, 303, 337
CS, 308
Cupric ion, 337
    as nutrient, 170
Cupric sulfate, 151
Cuprous ion, 337
Cuprous sulfide, 163
Cyclamate, 242
Cyclazocine, 295
Cyclohexane, 135
Cyclohexene, 7
Cyclohexylamine, 243
Cyclopropylmethyl group, 296
Cyclotron, 304, 337

p-cymene, 282
Cysteine, 171
Cytochrome, 172
Cytosine, 311

Dacron, 109
Dairy cattle, 75
DDE, 131, 138
    spectrum, 203
    structure, 136
DDT, 63, 105, 111, 129
    analysis, 195
    chemical name of, 133
    and eggshell thickness, 138
    in food chains, 132
    and gypsy moth, 148
    insect resistance to, 138
    in mammals' milk, 131
    and nervous systems of insects, 130
    o,p'-isomer, 129
    persistence of, 131
    production, 130
    solubility, 130
    spectrum of, 203
DDT-dehydrochlorinase, 138
Decision-making, and science, 302
Decomposers, 69
Defoliant, 307
Dendrite, 251
Dentistry, mercury in, 177
Deoxyribonucleic acid (see DNA)
Deoxyribose, 310
Department of Defense, U.S., 325
Depolarization, of neuron, 252
Depression, and barbiturates, 266
DES, 244
Detergents, 73, 80, 106
    synthetic, 86
Deuterium, 41
Development, 324
Dexedrine, 260
Diabetes, 320
Diacetylmorphine, 293
Dieldrin, 136
Diesel oil, 100
Diethylamine, 246
Diethylstilbestrol (see DES)
    and contraception, 229
Diglyceryl stearate, 238
Diimide, 65
Dimethyl disulfide, 121
Dimethyl sulfide, 121
Disorder, and chemical structure, 24
Distillation, 338
    of alcoholic beverages, 276
Dithiocarbamates, 152
DNA, 310
    backbone, 312
    chain, 313
    nucleotides in, 311
    of viruses, 319
Doctoral degree (see Ph.D. degree)
Dopamine, 255, 279
Double bond:
    in Δ'-THC, 284
Double helix, of DNA, 314
Drug industry, fungi control, 151
Dugway Proving Ground, 308
Dyes, 109

Ecosystems, 64
EDTA, 92
Education, and science, 302

Effects, of amphetamines, 261
  of barbiturates, 266
  of marijuana, 287
Efficiency, of glucose oxidation, 23
Eggshells, and DDT, 138
Einstein, Albert, 303
Elastomers, 106
Electric automobile, 52
Electric furnace, for steelmaking, 162
Electric potential, 251
Electric power resources, 32
Electrolysis, 338
  of aluminum, 166
  of copper, 164
Electromagnetic radiation, 338
Electromagnetic spectrum, 16
Electron, 39, 338
Electronic transitions, 338
Electrostatic precipitator, 61, 162, 187
Elm bark beetle, 145
Emulsifier, 238
Endothermic reaction, 22-23, 338
Energy consumption, and economic
  levels, 14
Energy crisis, 46
Energy, from fission, 303
Energy levels, 338
Energy resources, 33
Entropy, 24
Environment, 9
Environmental impact statement, 212
Environmental Protection Agency, 59,
  130, 140, 154, 159, 189, 210, 325
Environmental Science and Technology, 9
Enzymes, 138, 144, 275, 338
  synthesis of, 317
Epoxides, 148, 150, 338
Epoxy resins, 102, 108
Equilibrium, 25, 120, 338
  in aquatic ecosystem, 69
  proton transfer reactions, 90
Ergot, 270
Ergs, 41
Esters, 83, 89, 107, 144, 150, 339
  as food flavorings, 239
Estradiol, 8, 227-228
Estrogen, 224, 228, 245
Ethane, 27-28, 274
Ethanol (*see also* Ethyl alcohol), 83, 102
  111, 274
Ethanolamine, 254
Ethical problems, of scientists, 321
Ethyl acetate, 83, 264
Ethyl alcohol, 273, 274-280, 297
  use with barbiturates, 266
Ethylene, 54, 101, 107, 112, 115
Ethylenediaminetetracetic acid (*see* EDTA)
Ethylene dibromide, 51
Ethylene glycol, 102, 109
Ethylene oxide, 102, 109
17-α-ethynylestradiol, 228
Europe, population, 221
Eutrophication, 68, 91
Excited states, 18, 56, 339
Exothermic reactions, 23, 339
Extraction, 195

Fats, 238
Fatty acids, 86, 150
FDA, 240
  and DES, 245
  and nitrates, 246
Federal Radiation Council, 325
Fermentation, 99, 275

Fermi, Enrico, 303, 324
Ferric ion, 85, 339
  as nutrient, 169
Ferric oxide, 158
Ferrous metals, recycling of, 189
Fertilizers, 73, 75, 97
Fire ant, 141, 145
First Law of Thermodynamics, 24, 52
Fish, mercury in, 177
Fission, of uranium, 302
Flame ionization detector, 199
Flavorings, synthetic, 239
Fluidized bed reactor, 191
Follicle-stimulating hormone (*see* FSH)
Food additives, 237
Food Additives Amendment of 1958, 241
Food and Drug Administration (*see* FDA)
Food, Drug and Cosmetic Act of 1938,
  212, 240
Food industry, 98
Food packing, problems of, 237
Food process industries, 122
Forest resources, 115
Fossil fuels, 28
Fractional distillation, 99, 339
Free radicals, 47, 53, 339
  and food spoilage, 240
Frequency, of waves, 16, 339
Frost, Robert, 333
Fruit tortix, 147
FSH, 226
Fuel injection system, 52
Fuel oil, 100
Funding, of research, 323
Fungicides, 151
Furfural, 123
Fusion, 16
Future population, 219

GA, 307
Gamma rays, and cancer, 319
Gas chromatography, 197
Gaseous diffusion, 304
Gasification, of coal, 35
Gasoline, 100
Gelatin, 238
Genes, 310
Genetic code, 316
Genetic damage, and LSD, 272
  and marijuana, 287
Genetic diseases, 320
Genetics, 309
Geothermal energy resources, 33
Geraniol, 286
Glass, in solid waste, 187
Glucose, 18, 116, 275
Glutamic acid, 66
Glycerine, 83
Glycosidic linkage, 339
Gold, 165
Gonadotrophic hormones, 226
Gonads, 223
Government, and pollution control, 210
  and research, 325
Graduate education in science, 326
Grant proposals, 329
GRAS list, 241
  artificial sweeteners, 242
Ground state, 18
Groundwood process, 117
Growth, 318
Guanine, 311
Gypsy moth, 148

ΔH, 47
Haber process, 76
Hahn, Otto, 43, 302, 324
Hallucinogens, 268
  and amphetamines, 261
Hard water, 85
Harmaline, 268
Hashish, 281
Heat of reaction, 21
Helium, in gas chromatography, 197
Hematite, 158
Heme, 172, 174
Hemoglobin, 79, 246
  and lead poisoning, 173
Hemophilia, 309, 320
Hemp plant, 280
Hepatitis, 295
n-heptane, 102
Herbicides, 153, 207
Heroin, 293, 297
  addiction, 290
Hertz, 16, 340
Heterocyclic amines, 311
Hexane, 195
Hofmann, Arnold, 271
Hormones, 259
Human chorionic gonadotrophin, 227
Human wastes, 73
Hyaluronidase, 230
Hydrazine, 65
Hydrocarbons, 27
Hydroelectric resources, 32
Hydrogen, as energy source, 16
  in DNA, 314
  in Haber process, 77
Hydrogen bonding, 85, 116, 256, 274,
  340
Hydrogen chloride, 113
Hydrogen sulfide, 121, 171
Hydrogen sulfite, 120
Hydrologic cycle, 17
Hydrolysis, 90
Hydropulping, 189
Hydroxylamine, 65
Hydroxyl group, 8, 116, 307
7-hydroxy-THC, 288
Hypothalamus, 226

Imides, 264
In vivo conditions, 279, 340
Incineration, 187
  of food wastes, 123
  and polyvinyl chloride, 192
Industrial pollution, 74
Infrared radiation, 16, 340
Infrared spectrometry, analysis of auto
  exhaust, 206
Insecticides, alternatives to, 144
Insecticide, Fungicide and Rodenticide Act,
  212
Insects, larvae, 145
  metamorphosis, 144
  predators of, 145
  reproduction, 144
Insulin, 320
Intellect, chemistry of, 322
Internal combustion engine, 47, 49, 52
Intoxication, by alcohol, 276
Intrauterine device (see IUD)
Ion exchange resin, 86
Ionic bonds, 81
Ions, 81, 340

Iron, 158
  in aluminum ore, 166
  in Haber process, 77
  resources, 168
  in solid waste, 168
Iron cation, in hard water, 90
Isobutane, 101
Isobutene, 101
Isolation, of active marijuana agent, 281
Isopentane, 6
Isoprene, 108
Israel, 80
IUD, 232

Japanese beetle, 145
Juvenile hormones, 150

Kerosene, 100
Ketones, 54-55, 206, 234
Kilocalorie, 340
Kilowatt, 340
Knocking, 50
Kraft process, 119
Krebs cycle, 277

Lachrymators, 308
Lake Baikal, 72
Lake Erie, 64, 70
Lake Superior, 159
Lake Tahoe, 72
Land Grant College, 327
Larvae, 145
Latin America, population of, 221
Lattice structure, 340
Lawrence, Ernest, 304
Lead, 121
  analysis of, 205
  resources, 168
  toxicity of, 173
Lead additives, 50
Lead compounds, in auto exhaust, 206
Lead engine deposits, 50
Leavening agents, 239
Lecithin, 238
Leguminous plants, 66
Levulinic acid, 173
Lewis acid, 340
Lewis base, 92, 340
Lewis, G.N., 301
Lewis Structure, 6, 340
LH, 226
Ligands, 170, 340
Lignin, 116, 125
Lime, 94, 119, 162
  as fungicide, 151
Limestone, 25, 60, 119
  in steel manufacture, 159
Limiting Nutrient, 69
Limnology, 70, 340
Lindane, 135
Linear alkyl sulfates, 89
Lipids, 217, 341
  and DDT, 130
Liver, and alcohol, 277
Logarithm, 341
Long Island Sound, 75
Los Alamos, 305
Los Angeles, 187
  and smog, 58
Low-lead gasoline, 51
LSD, 297, 308, 271
Lubricants, 100

Lucite, 107
Luteinizing hormone (*see* LH)
Lysergic acid, 270
Lysergic acid amide, 271
Lysergic acid diethylamide (*see* LSD)

Magnesium ion, 85
  in chlorophyll, 172
  as nutrient, 169
Magnetic separation, of ferrous metals,
  158, 189
Malaria, 141, 150
Malathion, 142
Male oral contraceptive, 232
Malonic ester, 264
Maltose, 275
Mammals, and sex attractants, 148
Manhattan Project, 304, 324
MAO, 277
Marijuana, 280, 297
Mass spectrometry, 201
Master of Arts degree, 327
Master of Science degree, 327
Meats, nitrites in, 246
Medical uses, of marijuana, 288
Medicinal chemicals, 106
Mediterranean fruit fly, 145
Medroxyprogesterone acetate, 229-230
Membrane, of neuron, 251
Menstrual cycle, 226-227
Mercury, analysis of, 205
  in chlor-alkali process, 178
  spectrum of, 204
  toxicity, 176
Mercury cation, 92
Mercury-vapor lamp, 205
Mescaline, 268
Mestranol, 228
Metabolism, of alcohol, 276
Metal industries, 98
Metals, 157, 341
  consumption, 168
  resources, 167
  in solid waste, 187
  toxicity, 169
Meta substituent, 341
Methadone, 295
Methadrine, 260
Methane, 27, 187
Methemoglobinemia, 78
Methionine, 181
Methoxychlor, structure of, 136
Methylcyclohexane, 102
Methyl mercaptan, 121
Methylmercury, 179
Methyl methacrylate, 107
Methyl parathion, 142
Microbes, in oil slicks, 105
Microns, 341
Milk, 122, 238
Milk sugar, 294
Minamata Bay, Japan, 177
Mirex, 141
Mitochondria, 256
Molecular orbitals, 81
Monellin, 243
Money, and science, 302
Monitoring, of air pollution, 205
Monoamineoxidase, 256, 277
Monomers, 106
Monosodium glutamate, (*see* MSG)
Morning-after pill, 245
Morphine, 279, 290

Morphine receptor centers, 296
Mosquitos, 130, 150
Motor oil, 105
MSG, 243
Mueller, Paul, 120
Mumford, Lewis, 301
Mustard gas, 307
Mutation, 310, 318
Mycelles, 85, 341
Myoglobin, 172, 246
Myrcene, 147

Nabam, 152
NAD, 277
Naloxone, 295
Napalm, 307
Naphthalene, 103
Narcotic antagonists, 295
National Aeronautics and Space
  Administration, 325
National Bureau of Standards, 325
National Commission on Marijuana and
  Drug Abuse, 288
National Institutes of Health, 325
National Science Foundation, 325-6
Natural logarithm, 341
Natural resources, consumption, 169
Natural rubber, 108
Nembutal, 262
Nerve gas, 307
Nervous system, 249
Neuron membrane, 252
Neurons, 251
Neurotransmitter, 254
Nicotine adenine dinucleotide (*see also*
  NAD), 277
Nicotine, 267
Nitrate ion, 65
Nitrates, in fertilizer, 77
Nitrites, 65, 79, 246
Nitrogen, 64-5, 77
  in Lake Erie, 73
Nitrogen cycle, 67
Nitrogen oxides, 53
Nitrolotriacetic acid (*see* NTA)
Nitrosamines, 246
Nitrous acid, 246
Nixon, Richard, 308
NMR, 282
Noise Abatement and Control Act, 212
Nomenclature, 341
Non-bonding valence electrons, 7, 341
Noradrenaline, 255, 260, 277
Norepinephrine, 255
Norethindrone, 229-30
North America, population, 221
North American robin moth, 150
NTA, 91
Nuclear magnetic resonance (*see* NMR)
Nuclear reaction, 15, 43
Nuclear Test Ban Treaty, 305
Nucleic acids, 66, 68, 259
Nucleotides, 311
Nucleus, binding energy, 42
  of cells, 310
  diameter of, 40
Nutrients, in water, 64
Nylon, 6, 107, 265

Ocean dumping, of solid waste, 188
n-Octane, 47, 50
Octane, number, 102

Oil spills, 105
Oil tankers, 103
Olefins, 34, 206
Oleic acid, 84
Olevitol, 282, 286
Olive oil, 84
Olivitolic acid, 286
Open dump, 187
Open-hearth furnace, 161
Opiates, 290
Opium, 290
Oppenheimer, J. Robert, 305
Optical activity, 258
Oral contraceptive, 227
  and blood clotting problems, 231
  and cancer, 231
  long term, 233
  male, 232
Orbital energy level, 18
Orbitals, 341
Order in Chemical Reactions, 24
Organic structures, 5
Organic synthesis, 262, 342
Organization of research, 323
Organo-mercury compounds, as
    fungicides, 151
  toxicity, 180
Organophosphates, 142
Oriental foods, and MSG, 244
Orlon, 107
Ortho substituent, 342
Ovaries, 223
Oxidation, 21, 79, 277, 342
Oxidation potential, 342
Oxidation state, 54, 64, 342
Oxygen, and food spoilage, 237
Ozone, 55

Packaging, 192
Paints, lead containing, 176
Pairing, of nucleotides in DNA, 314
Palmitic acid, 84
Papaverine, 292
Paper, 116, 187
Paper chromatography, 197
Paper fibers, recycling of, 189
Paper industry, 98, 115
Para substituent, 342
Parasympathetic nervous system, 254
Parathion, 142
Parrot fever, 308
Particulate matter, from incinerators, 187
Particulates, analysis of, 205
  from steel, 162
Parts per million, 342
Pentachlorophenol, 152
Pentobarbital, 263
Perfumes, 99
Peripheral nervous system, 250
Peroxides, 112
Peroxy radical, 54
Peroxyacetylnitrate (PAN), 57
Peroxyacetyl radical, 56
Pervitin, 260
Pesticides, 106
  analysis of, 195
  benzene hexachlorides, 135
  and food supply, 127
Petrochemicals, 99
Petroleum, 99
  and coal industries, 98
Petroleum ether, 100
Petroleum industry, fungi control, 151
Petroleum refining, 99

Petroleum resources, 30
pH, 79, 91, 93, 118, 342
Pharmaceuticals, 99
Ph. D. degree, 326-7
Phenacyl chloride, 308
Phenobarbital, 262
Phenolic resins, 108
Phenols, 8, 111, 245, 343
Phenylalanine, 320
Phenyl group, 262
Phenylketonuria, 320
Phenylmercuric acetate, 151
Pheromones, 146
Phosgene, 306
Phosphate esters, 67
Phosphates, in detergents, 89
  in DNA, 310
  in fertilizer, 76
  in Lake Erie, 72
Phosphoric acid, 66, 68, 142, 238, 307
  acid-base equilibrium, 90
  as leavening agent, 239
Phosphorus:
  in pig iron, 161
Photochemical smog, 53
Photosynthesis, 18, 343
Phthalates, 114
pi (π)-Bond, 54
Pickling of steel, 162
Pig iron, 159
Pink bollworm moth, 147
Pituitary gland, 226
Plague, 153, 308
Plants, and chemical communication, 149
Plasticizers, 109
Plastics, in solid waste, 187
Plexiglass, 107
Plutonium-239, 304
Polar bonding, 81, 343
Pollution, and government responsibility,
    210
Pollution control equipment, 125
Polyacrylonitrile, 107
Polyamide, 107
Polychlorinated biphenyls, 113
Polyesters, 107
Polyethylene, 102, 107, 192
Polymer resins, 106
Polymers, 106, 116
Polymethyl methacrylate, 107
Polyolefins, 107, 343
Polypropylene, 102, 107
Polystyrene, 107, 112
Polyurethane, 108
Polyvinyl chloride, 107, 113, 192
Poppy, 290
Population problem, 217
Population, statistics, 219
Porphyrins, 172
Positron, 15
Postassium, chloride, 81
Potassium, in fertilizer, 76
Potassium, as nutrient, 169
  and neuron membrane, 251
Potential energy, 23, 343
Poultry, 75
Power generation, from solid waste, 191
Power requirements, in aluminum
    manufacture, 167
  in steel manufacture, 167
Prediction, of future population, 219
Preservatives, food, 239
President's science adviser, 325
Primary treatment of sewage, 94

Processing, of morphine, 294
Producers, in ecology, 69
Production, of iron and steel, 158
Progesterone, 225, 228, 230
Progestins, 225
Propylene, 107, 111
Prostaglandins, 233
  uterine response to, 235
Proteins, 68, 171, 259, 343
  as artificial sweeteners, 243
  structure, 343
Protons, 39, 343
Proton transfer reactions, 76, 84, 89,
  179, 239, 343
Psilocybin, 268
Psittocosis, 308
Psychotomimetic drugs, 270
Publication, of research, 329
Pulp and paper industry, 115
  fungi control, 151
Pure Food Act of 1906, 240
Purine, 311
Pyrimidine, 311

Quantum mechanics, 41, 343

Rabbit fever, 308
Radiant energy, 16
Radiation, and cancer, 319
RANN program, 326
Rat control, 153
Rats, learning in, 322
Recausticization, 119
Recycling, of solid waste, 188
Reduction, 27, 64, 79, 277, 343
  of iron ore, 159
Refuse, amount collected, 186
Release factors, 227
Rendering, of food wastes, 124
Reproduction, social influences upon, 236
Research, and EPA, 211
Research, funding of, 323
  in cancer, 319
  organization of, 323
Resolution, 343
Resolving optical isomers, 260
Respiration, 23
Retention time, 201
Rhythm method, 221
Ribonucleic acid (see RNA)
Ribose, 311, 316
Ribosome, 316
Rivers and Harbors Act, 211
RNA, 316
  in learning, 322
Roasting, of copper ore, 163
  of mercury ore, 178
Robin moth, 150
Rocky Mountain spotted fever, 308
Rodenticides, 153
Roosevelt, Franklin, 303
Rooster comb, 224
Rotary engine, 52
Rutherford, Sir Ernest, 39

Saccharin, 241
Salicylic acid, 110
Sanitary land fill, 187
San Joaquin Valley, 77
Saponification, 84
Sausages, nitrites in, 246

Schizophrenia, 323
Science, 9
Science, and ethical problems, 321
Scientific American, 9
Scientific method, 301
Scientists, and chemical warfare, 309
Scientists' Institute for Public Information, 9
Scopolamine, 268
Scotophobin, 322
Scrap metal, 168
Screwworm fly, 145
Sea of Galilee, 72
Secobarbital, 262
Seconal, 262
Secondary sewage treatment of food wastes,
  124
Secondary sex characteristics, 223
Secondary treatment of sewage, 94
Second Law of Thermodynamics, 46
Separation, of uranium isotopes, 304
Sequential birth control pill, 228
Serotonin, 255, 270, 277
Sewage, 91
Sewage treatment, 94, 211
Sewage waste, as fertilizer, 95
Sex attractants, insects, 146
Sex attractants, mammals, 148
Sex hormones, 8, 223
Sexual characteristics, 222
Sheep, 75
Sickle cell anemia, 309, 320
Sigma Xi, 9
Silver, 165
Silver ion, as Lewis acid, 171
Sinclair, Upton, 240
Singlet oxygen, 56
Singlet state, 344
Slag, 159-160
Smelting, 163
Smog, 53
Soaps, 80, 99
Soda ash, 92
Soda process, 117
Sodium benzoate, as food preservative,
  239
Sodium bicarbonate, 239
Sodium carbonate, 92
Sodium hydroxide, 119, 166, 178
Sodium ion, and neuron membrane, 251
Sodium nitrite, 246
Sodium p-laurylbenzenesulfonate, 86
Sodium polyphosphate, 89
Sodium pump, 251
Sodium silicate, 92
Sodium stearate, 82, 84
Sodium sulfide, 119
Soft drinks, 238
  artificial sweeteners in, 243
  packaging of, 193
Soft–hard acid–base theory, 181, 344
Solar energy resources, 33
Solid waste, composition, 187
  disposal costs, 191
Solubility, 344
Soviet Union, 305
  population of, 221
Speed of Light, 16
Spending, for chemical and biological warfare,
  308
Sperm, capacitation of, 230
Spermicides, 221

Stabilizers, 238
Starch, 20, 116, 275
Steam engine, 52
Stearate ion, 82
Stearic acid, 84
Stearyl alcohol, 89
Steel, 158
Sterilization, 221
  of insects, 145
Steroids, 223
Stockholm Conference, 213
Strip mining, 34
Structure, of enzymes, 317
  of $\Delta^{'}$−THC, 282
Strychnine, 268
Styrene, 103, 107, 108
Sulfamante, 242
Sulfate, 120
Sulfate process, 119
Sulfide, affinity for mercury, 181
Sulfite process, 120
Sulfonic acids, 86, 241
Sulfur, 120
  in coal, 35
  in coke, 160
  in pig iron, 161
Sulfur-containing gases, in sulfite process, 120
Sulfur dioxide, 60
  in copper smelting, 163
Sulfuric acid, 35, 60, 76, 129
Sulfur trioxide, 60
Superphosphate, 76
Supersonic transport, 5
Surfactants, 85, 344
Sveda, Michael, 242
Sweden, 220
Sweeteners, artifical, 241
Swine, 75
Sympathetic nervous system, 254
Synanon, 295
Synapses, 251, 293
Synergistic effects, of drugs, 261
  with marijuana, 288
Synthesis, of organic compounds, 262
  of $\Delta^{'}$-THC, 286
Synthetic rubber, 102
Szilard, Leo, 303

Taconite, 158
Tear gas, 308
Technology, 4, 344
Tenure, 329
Teratogenic effects, 92
Terpenes, 147, 286
Terphthalic acid, 109
Tertiary treatment of sewage, 95
Testes, 223
Testosterone, 223
2, 3, 7, 8-Tetrachlorodibenzo-p-dioxin, 154
Tetraethyl lead, 50, 175
Tetrahedral carbon, 258
$\Delta^{'}$-tetrahydrocannabinol (*see* $\Delta^{'}$-THC)
Tetrahydropapaveroline, 279
Tetravalency of carbon, 6, 344
Textile industry, 98
  fungi control, 151
Textile wastes, 187
Thallium, toxicity of, 184
$\Delta^{'}$-THC, structure of, 284
Thermal conductivity detector, 199
Thermal pollution, 44, 46

Thermodynamics, 23
Thermoplastic polymers, 106
Thermosetting polymers, 106
Thickeners, 238
Thompson, J.J., 39
Thymine, 311, 316
Tin, resources, 168
Tolerance, to barbiturates, 266
Toluene, 51, 102, 103
Torrey Canyon, 104
Toxaphenes, 136
Toxicity, of beryllium, 183
Trace minerals, 170
Trans geometry, 233, 282
Treatment, of cancer, 319
Triglyceral stearate, 82
2, 2, 3-Trimethylpentane, 50
Triplet state, 56, 344
Truman, Harry, 305
Tularemia, 308

Ultraviolet radiation, 16, 344
Ultraviolet spectrum, 201
United Nations, 213
United States Department of Agriculture, 325
United States Weather Service, 325
Unsaturation, 344
Uranium, 43, 302
Uranium-235, 302
Uranium committee, 303
Uranium hexafluoride, 304
Uracil, 316
Urea, 264
Use, of amphetamines and barbiturates, 266
USSR, population, 221
Uterus, response to prostaglandins, 235

Vanadium pentoxide, 60
Vasectomy, 221, 232
Venezuela, 103
Venturi scrubber, 162
(−)-verbenol, 286
Veronal, 262
Vibrational transitions, of molecules, 206
Vietnam, 307
  herbicide use in, 154
Viral agents, 308
Viruses, as insect predators, 145
  and cancer, 319
Visible light, 344
Visible radiation, 16
Visible spectrum, 201
Vitamins, as food additives, 240
Volatility, 99, 344
Voltage in nerve, 251

Wald, George, 248
Wankel engine, 52
Warfarin, 153
Waste paper, 116
Water, as Lewis base, 171
Water Pollution Control Act, 211
Water pollution, in food process industries, 122
  sulfite process, 121
Water softening, 86, 91
Wavelength, 16, 344
Waves, 16
Waxes, 100

Withdrawal effects, of heroin, 294
  from amphetamines, 261
Wood, 116
Wood, John M., 180
World War I, 306

X-ray analysis, of enzymes, 317
X-rays, and cancer, 319

o-xylene, 103
p-xylene, 103, 109

Yeast, 275

Zinc ion, in carboxypeptidase, 173
Zinc, resources, 168

Zymase, 276

0 3 7 0